Springer Series in Fashion Business

Series editor

Tsan-Ming Choi, The Hong Kong Polytechnic University, Hung Hom, Hong Kong

More information about this series at http://www.springer.com/series/15202

Yingjiao Xu · Ting Chi · Jin Su
Editors

Chinese Consumers and the Fashion Market

Springer

Editors
Yingjiao Xu
North Carolina State University
Raleigh, NC
USA

Jin Su
The University of North Carolina
Greensboro, NC
USA

Ting Chi
Washington State University
Pullman, WA
USA

ISSN 2366-8776 ISSN 2366-8784 (electronic)
Springer Series in Fashion Business
ISBN 978-981-10-8428-7 ISBN 978-981-10-8429-4 (eBook)
https://doi.org/10.1007/978-981-10-8429-4

Library of Congress Control Number: 2018931923

Printed on acid-free paper

This Springer imprint is published by Springer Nature
The registered company is Springer Nature Singapore Pte Ltd.
The registered company address is: 152 Beach Road, #21-01/04 Gateway East, Singapore 189721, Singapore

Preface

China has experienced a great economic boom in the past two decades, leading to a very dynamic consumer market in the country. Urbanization is happening at a historical high rate. The post-80s one-child generation has been entering the workforce. A larger percentage of the huge Chinese population has joined the middle class, with significantly increasing purchasing power. The upper middle class and the globally minded young consumers are exerting disproportionate influence in the market. Significant shifts have been witnessed in consumption dynamics in the Chinese consumer market.

Fashion products, with the symbolic and social communication function, have been playing a significant role in Chinese consumers' lives. Deeply rooted in the Chinese culture, face consciousness plays an important role in consumers' fashion consumption. Chinese consumers have been strongly associated with a high level of brand consciousness and brand loyalty. The recent years have witnessed an explosion of Chinese consumers' craving for luxury fashion products. Chinese consumers have been found to hold a discriminant preference of global brands over Chinese domestic brands. With the fact that many global brands manufactured in China and at the same time Chinese companies shifting from the export orientation to focusing on branding in the domestic market, will Chinese consumers' pursuit of global brands shift accordingly? Winning the Chinese consumers is rewarding but challenging for both global and domestic brands.

The consumption dynamics is also spurred by the advancement of technology in product design and production, marketing, and retailing. Online shopping, particularly mobile shopping, has brought revolutionary changes to the retail industry in China. E-tailers such as Taobao, T-mall, and several others have become the main providers of young consumers' daily needs, including fashion products. However, coming together with the convenience and low prices provided by online shopping is the concern of low quality and counterfeits. However, this challenge facing online shopping provides opportunities for the in-store retailers to attract and retain their customers. Another major implication of technology advancement in the Chinese fashion market is consumers' brand behavior via social media. Online product reviews, brand community, new shopping platforms enabled by social media such as the most powerful WeChat as well as group shopping all bring additional dynamics to the Chinese consumer market.

Changes in consumers' lifestyles and values leave a strong impact in the Chinese consumer market. The severe pollution in the country not only stimulates a high level of consciousness of health and accordingly exercise, but also brings awareness and concern for environmental protection. Consumption of sportswear, athleisure apparel and footwear products, green products, as well as secondhand clothing attracts great attention from the marketers. Another impact is from the real estate market. As housing getting extremely expensive in China, young adults either become very thrift in spending to save for future housing needs or choose to give up the motivation to buy a house but instead focus on current gratification.

Lastly, the dynamics of the Chinese market is also manifested in the form of bipolar coexistence. Globalization has produced a hybrid of modern and traditional consumers coexisting in the Chinese fashion market. Also coexisting in the Chinese market are the mass low-income consumer segment and the superrich. Imbalance is observed also between consumers living in the inner land and those in the coastal.

With no intention to cover all the aforementioned aspects, this edited book aimed to provide an insight into consumer behaviors in the growing and dynamic Chinese fashion market. We hope this book can provide an inspiration to peer scholars and practitioners in their endeavors to investigate and understand the Chinese fashion market.

Raleigh, USA Yingjiao Xu
Pullman, USA Ting Chi
Greensboro, USA Jin Su

Contents

Part I
Apparel Customization and Luxury Fashion in China

Facing the Rising Consumer Sophistication: Identifying the Factors that Influence Chinese Consumers' Intention to Purchase Customized Apparel

Chunmin Lang, Ruirui Zhang and Li Zhao

Abstract Consumers' desire for novelty and uniqueness drives consumers to consistently seek new products that differentiate them from others. New goods are purchased frequently and disposed of quickly by consumers, which provide a short-term experience of novelty and newness but increase landfill wastes. Apparel customization provides consumers with an option to create their own personalized clothing items, which can increase product longevity because it results in personal attachment from customized styles and well-fitted garments. The study investigated the influence of shopping motivations and perceived enjoyment on consumer intention to purchase customized apparel for Chinese consumers. By applying the theory of planned behavior (TPB), this study considers that engaging in customization is not only from an individual's values and self-interest, but also from social values. An online survey was conducted with 321 Chinese consumers. The SEM analysis results indicate the three shopping values (adventure shopping, idea shopping, and achievement shopping) significantly influence the intention to purchase customized apparel products in an indirect way, and the significant effect of perceived enjoyment on the intention toward apparel customization is also confirmed. Further, the mediating effects of TPB components were also confirmed among Chinese participants. This study highlights the role of shopping values and perceived enjoyment on the adoption of apparel customization and provides guidance for the retailers and marketers who provide apparel customization services in the Chinese market.

C. Lang (✉)
Department of Textiles, Apparel Design, & Merchandising, College of Agriculture, Louisiana State University, 143 Human Ecology, Baton Rouge, LA 70803, USA
e-mail: cmlang@lsu.edu

R. Zhang
Department of Fashion Design and Retailing, Framingham State University, 234A Hemenway Annex, Framingham, MA 01701, USA
e-mail: rzhang@framingham.edu

L. Zhao
Textile and Apparel Management, University of Missouri, 137 Stanley Hall, Columbia, MO 65211, USA
e-mail: zhaol1@missouri.edu

Y. Xu et al. (eds.), *Chinese Consumers and the Fashion Market*, Springer Series in Fashion Business, https://doi.org/10.1007/978-981-10-8429-4_1

Keywords Apparel customization · Chinese consumers · Shopping motivations Perceived enjoyment · Theory of planned behavior

Introduction

Notably, consumers' desire for novelty and uniqueness has a great impact on the economic vividness of the fashion industry (Kawamura, 2005). These needs drive consumers to constantly seek new products that differentiate them from others. New goods are purchased frequently and disposed of quickly by consumers, which provide a short-term experience of novelty and newness (Birtwistle & Moore, 2007), but increase landfill waste. This is the case especially in China, the country with the largest population in the world, which produces much of the material that ends up in landfills. Approximately 26 million tons of textile waste is generated in China annually, which makes up about 3.5–4% of the world's waste materials (Ravenhall, 2012; Xue, Zhang, & Wang, 2014). Pressure on the environment requires a radical change among consumers and the fashion industry to develop creative solutions in garment production and consumption (Lawless & Medvedev, 2016). Apparel customization provides consumers with an option to create their own personalized clothing items, which can increase product longevity because it results in personal attachment to customized styles and well-fitted garments (Armstrong, Niinimäki, Kujala, Karell, & Lang, 2015; Flynn & Vencat, 2012). Participating in the process will create a deeper emotional connection between the consumer and the product that he/she creates (Lee & DeLong, 2017); thus, the product may be kept for a longer period of time (Armstrong et al., 2015). Furthermore, this creative and innovative marketing campaign has also become a significant marketing strategy to develop and maintain loyal customers (Cho & Fiorito, 2009).

Although apparel customization may hold substantial benefits for businesses and societies and is continuing to expand and gain more attention, not all consumers have been engaged in the transactions (Boer, Pedrazzoli, Bettoni, & Sorlini, 2013; Sherman, 2014), especially in China (Kesari & Atulkar, 2016; Ou, 2011), which indicates the need to understand what influences those consumers' intention to purchase customized apparel products. Shopping values are critical psychological variables that encourage consumers to behave in a particular manner (Solomon & Rabolt, 2006). They affect the way consumers respond to different market strategies and how they consume particular products or services. Despite the growing practical importance of apparel customization, there is a lack of knowledge in the effect of shopping values on consumers' attitudes and intentions of purchasing customized apparel products (Cho & Fiorito, 2009). Therefore, studying how shopping values influence the engagement in apparel customization may aid in identifying approaches to increase consumers' purchase intention of customized apparel.

The majority of previous research has been focused on customization strategy, technology, supply chain management, and operation systems (McKinney, Gill,

Dorie, & Roth, 2017; Sanders & Stappers, 2015; Senanayake & Little, 2010). In regard to consumers' perception of apparel customization, some exploratory qualitative studies have focused on consumer preferences and motivations, but most of the research was conducted in the US or European countries. For example, Armstrong et al. (2015) conducted focus groups in the US and Finland to uncover consumers' preferences in regard to the participatory design of apparel products. Niinimäki and Hassi (2011) conducted a qualitative study with two online questionnaires in Finland to investigate consumers' interest in co-creation of textile products. Challenges to adoption, such as uncertainty about personal style (Armstrong et al., 2015), unfamiliarity with technology involved in the customization process (Lee & Chang, 2011), concerns of waiting for a long time (Armstrong et al., 2015; Kim & Hong, 2011), and lack of trust in the provider (Armstrong et al., 2015; Ou, 2011) have been identified previously. For instance, some consumers were concerned about the lack of guarantee of a successful result if they participate in the design process. Some quantitative studies conducted in the US have confirmed the positive influences of attitude, ease of use, perceived enjoyment, exciting experience, as well as perceived security for mass customization (Cho & Fiorito, 2009; Fiore, Lee, & Kunz, 2004; Lee & Chang, 2011). But studies focusing on Chinese consumers' apparel customization behavior are lacking. Ou (2011) carried out in-depth interviews to explore young Chinese consumers' perceptions of mass apparel customization. However, to date, the role of shopping values in the adoption of customized apparel for Chinese consumers has not been investigated.

McKinsey & Company's report (Atsmon, Magni, Li, & Liao, 2012) suggested that China's consumer sophistication is on the rise. By the year of 2020, half of China's urban households' income will catapult into the upper middle class. As income level rises, Chinese consumers tend to be more rational about fashion and begin to pay more attention to clothing that reflects their personal style (Atsmon et al., 2012). These consumers have indicated higher demand for innovative and personalized products. These rising expectations among Chinese consumers require the market to provide products with good quality and the ability to cater to their individualized needs. However, many fashion companies still do not have adequate capabilities to process the customized products to meet consumers' personalized requirements due to lack of understanding of these consumers. Although a variety of studies have been done on customization in developed countries (Kim & Hong, 2011; Lee & Chang, 2011; Niinimäki & Hassi, 2011; Sanders & Stappers, 2015; Senanayake & Little, 2010), understanding of Chinese consumers' motivations toward apparel customization is lacking.

The purpose of this study is to determine the factors that influence Chinese consumers' intention to purchase customized apparel. The study results are expected to contribute to the broader literature that focuses on customization in the apparel industry and to lay the foundation for the development of strategies that encourage apparel customization. In addition, by applying Ajzen's (1991) theory of planned behavior (TPB), this study considers that engaging in customization is not only of an individual's values and self-interest, such as attitude and perceived behavior control, but also of social values, like subjective norms. Specifically, personal values are considered to be external variables and assumed to influence intention indirectly mediated by the components of TPB (Ajzen & Fishbein, 1980). This study explores the influence of shopping values and perceived enjoyment on the components of TPB, as well as consumers' intention to purchase customized apparel. This research is the first attempt to examine the relationship of shopping values and the intention to purchase customized apparel. The results of this research will provide managerial insights for the promotion of apparel customization.

Literature Review and Hypotheses Development

Apparel Customization

Customization is defined as delivering products and services with enough variety and uniqueness that best meet an individual's needs (Cho & Fiorito, 2009; Pine, 1993). Product customization assumes that the market is heterogeneous and consumer-driven. It highlights a consumer-centered design approach. In order to meet the consumer's specific requirements effectively, consumers' involvement during the product development cycle is crucial. The essence of customization is to transform a consumer into a co-designer (Tseng & Piller, 2003). Co-design is a collaborative process between the consumer and the retailer, through which the consumer gets involved in the design process and a product is customized to fulfill the consumer's individual requirements (Peterson, 2016). Therefore, to enhance the interaction with consumers, the co-design process needs to trigger the hedonic motivation among consumers and improve the enjoyment of the process (Piller & Tseng, 2010).

Apparel customization is an important competitive advantage in the fashion industry (Salvador, De Holan, & Piller, 2009). It has become a significant marketing strategy in the development and maintenance of loyal customers and to fulfill consumers' demands more precisely (Steen, Manschot, & Koning, 2011), therefore, to increase consumers' satisfaction with the product. In addition, Pine (1993) also described customization as an effective approach to eliminate excessive inventory, since products are made with consumers' specifications, therefore, to reduce waste. Consumers also benefit from customization. Involvement in customization allows consumers to specify unique requirements and apply personal preferences and uniqueness into the design (Ou, 2011; Schreier, 2006). In addition, the custom design provides the consumer with increased product satisfaction by designing a product that is more personalized in fit, style, comfort, as well as functionality (Piller & Müller, 2004; Schreier, 2006). Consumers also experience entertainment and enjoyment during the process of co-design (Schreier, 2006; Steen et al., 2011) and the pride of authorship (Schreier, 2006). Where environmental sustainability is concerned, product customization may bring great benefits as well, such as maximizing usage by increasing consumer satisfaction and improving the fit between the product's properties and the consumer's needs (Bruno, Nielsen, Taps, & Jorgensen, 2013). Customization is considered as environmentally sustainable (Medini, Da Cunha, & Bernard, 2015). To make customized products, efforts and pride have been involved in the process, and the process of customization might be an unforgettable experience; therefore, consumers tend to build a solid emotional bond with the customized products and keep the items for a longer time. Consumers' involvement in the design process can result in a deeper relationship and personal attachment to the garment, due to customized styles and well-fitted pieces, therefore, increasing the use frequency and potential life span of the product and a reduction in landfill waste (Armstrong et al., 2015; Chapman, 2009; Niinimäki & Hassi, 2011).

Short use time of garments is related to dissatisfaction that can be caused by low quality and poor fit (Armstrong et al., 2015; Niinimäki, 2014). To improve the fit and to meet individual specifications and requirements, customized apparel has become increasingly popular among consumers (Senanayake & Little, 2010). The apparel industry has built popular customization services, which allow consumers to participate in product development in different ways. For example, consumers can get involved in the design stage by selecting colors, fabrics, trims, and silhouette (Kamali & Loker, 2002). Some prominent examples that exist in the industry include: NikeiD studio, a digital customization platform that provides services for customers to configure their own sneakers by choosing different colors and materials (Abnett, 2015); Blank Label, an online male clothing customization retailer that offers a variety options of fabrics, styles, and colors for consumers to build the apparel that looks and fits the way they want; eShakti provides consumers with options of different styles and designs to customize their dresses, skirts, and tops for better fits; Vans provides consumers the chance to design their own skate shoes by combining any colors or patterns for the different parts of the shoes. Besides participating in the co-design, consumers can also participate in pattern making to get custom fit clothing (Fiore, Lee, & Kunz, 2003); Nomo Jeans, for example, is a fashion company that utilizes 3D body scanning to create made-to-measure jeans and to improve personal fit. In the process of customization, the consumer is able to apply their own creativity, preferences, narratives, and even memories to the product design (Armstrong et al., 2015). It provides consumers with a choice of obtaining unique and better fitting products.

China has become a profitable apparel market due to its large population and rapid economic growth (Atsmon et al., 2012). Despite the growing importance of apparel customization and the development of the Chinese market, the concept of customization is rather new to most people in China and not many people have been engaged in the transaction (Ou, 2011), and the acceptance of apparel customization among Chinese consumers is still low (Kesari & Atulkar, 2016). Therefore, there is a critical need to examine the personal factors influencing the intention to purchase customized apparel among Chinese consumers.

Shopping Values

Shopping values are critical psychological variables that encourage consumers to behave in a certain manner (Solomon & Rabolt, 2006). Consumers are motivated not only by the utilitarian value, but also the feeling obtained during the shopping process. Shopping values are mainly categorized as utilitarian and hedonic drives (Arnold & Reynolds, 2003). Two dimensions, achievement shopping and efficiency shopping, were included in utilitarian shopping values (Kim, 2006). Efficiency shopping value states that consumers want to be time-savers during the shopping process, while achievement shopping value refers to task-oriented shopping motivation where getting things done is critical (Kim, 2006). Hedonic shopping value contains six dimensions, which reflects the joy, entertainment, and pleasure derived from the experience (Arnold & Reynolds, 2003). Among the six dimensions, adventure shopping refers to shopping for stimulation, adventure, excitement, and the feeling of being in another world. Social shopping reveals shopping as a part of consumers' social life. Gratification views shopping as a special treat for consumers to alleviate a negative mood for stress relief. Idea shopping refers to the process of consumers' trying to gather information, to keep up with the new fashion trends, and to see new products with innovations. Role shopping reflects the enjoyment and excitement that consumers feel when shopping for others. Value shopping emphasizes the joy of looking for discounted projects and the best deals.

A previous study conducted by Merle, Chandon, Roux, and Alizon (2010) has acknowledged that customization provides both utilitarian and hedonic benefits. According to Xu and McGehee (2012), Chinese consumers pursue novelty and new trends during their shopping experience to satisfy hedonic value. The co-design process in customization, in itself, conveys adventure and provides fun and enjoyable experiences (Ou, 2011). For some consumers, getting involved in the process is also an approach to obtain new fashion skills, and those consumers may feel fulfillment and pleasure by wearing clothing they created themselves (Ou, 2011). Therefore, by carefully reviewing the features of each dimension of both utilitarian and hedonic shopping values, as well as the characteristics of apparel customization, only adventure shopping value, idea shopping value, and achievement shopping value were investigated in the context of purchase intention toward apparel customization among Chinese consumers. These three shopping values are believed to be highly relevant to the experience of apparel customization and having the greatest potential for providing insights into consumer intentions to purchase customized apparel. As personal values, these variables are taken as external factors and assumed to influence intention indirectly through the components of TPB (Ajzen & Fishbein, 1980).

Adventure shopping refers to shopping for the purpose of experiencing excitement and adventure (Arnold & Reynolds, 2003) and describes how consumers seek freshness and need stimulations in order to escape from the feeling of boredom (McGuire, 1974). Apparel customization provides consumers with visual stimulation and exciting experiences in the design and fashion-making process. In addition, customized services could offer unique and fresh suggestions to consumers. Hence, consumers may enjoy shopping in such an atmosphere and develop a positive attitude toward the process of customization. A previous study has confirmed that hedonic shopping values have an influence on attitude (Chung, 2015). Furthermore, consumers with adventure shopping values always seek excitement and adventure in their shopping process and they may establish strong positive beliefs about their ability of apparel customization, for which special skills may be needed. In addition, a previous study has confirmed the significant positive relationship between the preference of having an exciting experience and the willingness to accept mass customization (Fiore et al., 2004). In the context of Chinese consumers, a previous study has found that adventure shopping value positively influences purchase intention among Chinese consumers (Xu, Chen, Burman, & Zhao, 2014). Lau, Sin, and Chan (2005) have also confirmed that Chinese consumers' cross-border shopping behavior is positively impacted by adventure shopping. Therefore, the following hypothesis was proposed:

H1: Adventure shopping value positively influences (a) attitude toward, (b) subjective norms of, (c) perceived behavior control of, and (d) perceived enjoyment of apparel customization.

Idea shopping refers to shopping to gather information about new trends, fashion, and products (Kim & Hong, 2011), in order to update the knowledge of new style and innovations (Arnold & Reynolds, 2003). Consumers' gathering of information for a specific purchase need represents a leisure pursuit as a final goal. Nowadays, consumers, especially younger generations, constantly seek new ideas and fashion items to fuel their desire for novelty and uniqueness. Taking part in customization for fashion products could provide a means for those consumers who are seeking unique, stylish, and fashionable items. Getting involved in customization to make one's own one-of-a-kind fashion product might be an exciting approach to gain ideas about new fashion and show their pioneering role in the history of acceptance of new ideas. Therefore, they may have a more positive attitude toward apparel customization and tend to enjoy the process of co-design. Additionally, the activity of apparel customization can be seen as a way of collecting information of creativity. People with idea shopping values are usually fashion leaders (Kim & Hong, 2011), who tend to adopt new trends earlier than others; they would be more confident about their creativity and enjoy the experience of the customization process. Therefore, it is hypothesized that:

H2: Idea shopping value positively influences (a) attitude toward, (b) subjective norms of, (c) perceived behavior control of, and (d) perceived enjoyment of apparel customization.

Achievement shopping refers to a goal related shopping experience where consumers feel is important to accomplish what they had planned on a particular shopping trip (Kim, 2006). As a utilitarian value, achievement shopping value emphasizes the efficiency and functional benefits obtained during the process

(Irani & Hanzaee, 2011). Product specifications and quality of customized products can fulfill consumers' basic needs and assist consumers in enjoying the decision-making process (Kesari & Atulkar, 2016; To, Liao, & Lin, 2007). Utilitarian has been perceived as one benefit for customized products (Merle et al., 2010), which integrates both aesthetic and functional fit (Schreier, 2006). Franke and Schreier (2008) have confirmed that a positive relationship exists between the utilitarian benefit and the willingness to pay more for customized products. Through apparel customization, consumers' specific need for a unique style or better fit can be achieved. For instance, for consumers who have some special needs, such as plus-size clothing, it is crucial to accomplishing their shopping goals of obtaining better fit through apparel customization. By participating in the process of customizing products, consumers may also feel a sense of accomplishment. Therefore, they may enjoy more throughout the process of co-design. Furthermore, individuals with achievement shopping value are usually functional shoppers (Kim, 2006); they tend to be goal-oriented and have strong beliefs about their ability and are concerned about others' opinion on their achievement. In addition, achievement value has also been found to positively influence consumers' attitude (Jayawardhena, 2004). Thus, the following hypothesis is proposed:

H3: Achievement shopping value positively influences (a) attitude toward, (b) subjective norms of, (c) perceived behavior control of, and (d) perceived enjoyment of apparel customization.

Perceived Enjoyment

Perceived enjoyment is an intrinsic motivation (Teo, Lim, & Lai, 1999) and interprets the extent to which performing a certain activity is perceived to be enjoyable in itself, regardless of the performance consequences (Davis, Bagozzi, & Warshaw, 1992). As the intrinsic motivation factor, perceived enjoyment indicates the pleasure and satisfaction gained from performing a certain behavior. It is a salient factor affecting consumer attitudes and intentions toward the behavior. Attitude represents an individual's general evaluation of performing a certain behavior (Ajzen, 1991), while perceived enjoyment focuses on how fun the process of apparel customization is. The positive influence of perceived enjoyment on attitude has been confirmed by several studies (Childers, Carr, Peck, & Carson, 2001; Lee & Chang, 2011), indicating that when the activity is perceived to be enjoyable, a positive attitude toward the behavior is developed. In addition, consumers who engage in designing their own products will experience positive or negative emotions during the interaction. When consumers enjoy and feel interested in the co-design process of apparel customization, they may perceive positive beliefs about their own ability and develop confidence in the control of behavior throughout the interaction.

Enjoying the process is an important motivator for people to engage in a certain activity. Individuals may get involved in a particular behavior if it yields fun and pleasure. The co-design process in apparel customization delivers interactive features, which requires consumers to communicate and interact with the professional designer, retailer, or technological tool. Especially for apparel products, in order to achieve the particular style they anticipate, consumers may need to try different choices provided by retailers. Not feeling comfortable and enjoying the process may discourage their intention. The positive influence of perceived enjoyment on mass customization has been confirmed by a variety of research (Franke & Schreier, 2010; Lee & Chang, 2011). Given that co-design process of apparel products is a form of a hedonic system, perceived enjoyment is expected to have a salient influence on purchase intention toward apparel customization. Therefore, the following hypotheses were generated:

H4: Perceived enjoyment of apparel customization is positively associated with (a) attitude, (b) subjective norms, and (c) perceived behavior control.
H5: Perceived enjoyment of apparel customization is positively associated with consumers' intention to purchase customized apparel products.

The Theory of Planned Behavior

The framework of the theory of planned behavior (TPB) indicates that an individual's intention toward a certain behavior is affected by attitude, subjective norms, and perceived behavior control. Attitude represents an individual's evaluation of performing certain behaviors and is regarded as a major determinant of the behavior (Ajzen, 1991). A positive attitude toward a given behavior indicates that an individual is more likely to participate in the behavior. For example, research has shown that consumers' attitudes toward mass customization positively influence their responses, which includes the intention to purchase and to recommend customized products to others (Lee & Chang, 2011). The customization process contains interactive features between consumers and retailers; therefore, it would be crucial to examine consumers' attitudes and its effects on their responses.

Subjective norms refer to the perception of how important reference groups' expectations of exhibiting a certain behavior are. An individual's behavior is not only influenced by personal values, but it is also influenced by social groups, especially by specific reference groups. The opinions regarding certain behaviors from those people who are important to an individual play a crucial role in consumers' response to the behavior, which was also confirmed (Salazar, Oerlemans, & van Stroe Biezen, 2013). China especially has been identified as a country with a collectivist culture (Hofstede, 1980), which indicates that family, friends, and community play an important role in an individual's attitude toward a certain behavior.

Perceived behavioral control refers to an individual's perception of his or her ability to perform a certain behavior. Previous research has illustrated the positive influence of perceived behavior control on people's feelings about the behavior and their expected outcomes (Lee & Chang, 2011). The co-design process of apparel customization requires the consumer to have certain skills and ideas regarding the style they expect and to interact and communicate with the retailer or designer. The perception of inability to have control over the co-design process may discourage people's intention toward participating in apparel customization (Piller, Schubert, Koch, & Moslein, 2005). A strong perception of the ability to perform co-design points to a higher intention. Therefore, the research hypotheses are stated below:

H6: Consumers' (a) attitude, (b) subjective norms, (c) perceived behavior control positively influence consumers' intention to purchase customized apparel products.

Research Method

Sample and Data Collection

Data were collected in China with a convenient sampling strategy. An online survey was set up on a Chinese online market research Web site called *Wenjuanxing*. Participants were recruited via *WeChat*, which is a popular Chinese social media app having over 889 million active users in 2016 (CIW Team, 2016). Three researchers sent the invitation with the survey link to their contacts through *WeChat* and requested them to complete the survey, as well as to share the survey link with their contacts on *WeChat*. In total, CNY 240.00 (US$36.42) *Red Envelope* was distributed to the contacts; individuals who completed the survey received a random amount as an incentive. Each respondent was given a brief description of the study, explaining Institutional Review Board approval and the implications of his or her participation. In total, the survey link was sent to 878 WeChat users from September to November 2016. A total of 321 usable responses were collected and used in the data analysis, with a response rate of 36.6%.

Instrument Development

Three sections were included in the questionnaire. The first section contains a series of multi-item scales to measure the constructs including adventure shopping value, idea shopping value, achievement shopping value, perceived enjoyment, attitude, subjective norms, and perceived behavior control. All of these constructs were measured on five-point Likert scales with 1 = strongly disagree to 5 = strongly agree. The items used to measure adventure shopping value, idea shopping value, and achievement shopping value were borrowed from Arnold and Reynolds (2003) and Kim (2006). Perceived enjoyment was measured by five items modified from Merle et al. (2010). A semantic differential scale was employed to measure attitude (Ajzen, 2002). An incomplete statement regarding apparel customization was developed, followed by five groups of semantic differential adjectives. Based on the five-point scale, participants were requested to select the corresponding adjective from each group to complete the statement. Subjective norms were measured using a three-item scale modified from Ajzen (2002). Perceived behavioral control was measured by four items modified from Kidwell and Jewell (2003).

The second section measures consumers' purchase intention toward apparel customization. To better explain the idea of apparel customization, one hypothetical scenario statement was developed based on existing industry customization examples. Participants were requested to rate the level of willingness to purchase customized apparel products by responding to the question, "I intend to purchase customized apparel products during the next 12 months?" The scenario was described as following:

> A fashion retailer provides consumers with customization services: consumers can choose their favorite style, color, fabric, pattern, size, and image, etc. and design their favorite, one-of-kind style of clothing with the assistance of professional designers and place the order. The retailer will make the clothing according to consumers' design; in the meantime, this retailer also provides relative consumer services as needed.

The third section measures consumers' demographics. The questionnaire designed for this study was originally developed in English because all the instruments adopted in this study were written in English. Then, the questionnaire was first translated into Chinese by one bilingual scholar who has a Ph.D. degree in apparel merchandising, and then the Chinese version was back-translated to English by a different bilingual scholar to ensure an accurate translation.

Results

Profile of the Participants

Of the total 321 participants, the majority of them (63.9%) were female. Participants aged 18–35 made up 67% of the sample, followed by 36–50 with 27.7%. Almost half (49.2%) of the participants had a college degree, and there were also 29.9% participants reported having a master's/MBA or higher degree. With regard to annual household income, 31.2% of the participants reported annual income being more than CNY 200,000 (US$30,367); 31.4 and 47.4% of participants had annual income between CNY 100,000 (US$15,184) and CNY 199,999 (US$30,366) and less than CNY 99,999 (US$15,183), respectively. Table 1 presents the demographic summary of participants.

Table 1 Demographic summary of participants (n = 321)

Consumer profile	Frequency	Percentage
Gender		
Male	116	36.1
Female	205	63.9
Age		
18–35	215	67.0
36–50	89	27.7
51–69	17	5.3
Highest education		
High school graduate or lower	24	7.5
Associate degree	43	13.4
College graduate	158	49.2
Master/MBA or higher	96	29.9
Annual household income		
Less than CNY 69,999 (US$10,628)	74	23.1
CNY 70,000 (US$10,629)–CNY 99,999 (US$15,183)	78	24.3
CNY 100,000 (US$15,184)–CNY 169,999 (US$25,812)	73	22.7
CNY 170,000 (US25,813)–CNY 199,999 (US$30,366)	28	8.7
CNY 200,000 (US$30,367) or above	68	21.2

Measurement Model

Reliability and validity were both examined in this study. A two-step approach was adopted (Anderson & Gerbing, 1988). Confirmatory factor analysis (CFA) was first conducted to confirm a model for the measurement of each latent variable (Jackson, Gillaspy, & Purc-Stephenson, 2009). Then structural equation modeling (SEM) was employed to test the proposed hypotheses.

The result of the confirmatory factor analysis of the measurement model, including seven latent constructs with a total of 26 items, exhibited a good model fit ($\chi^2_{(df=302)}$ = 845.637, $p < 0.000$, $\chi^2/df = 2.80$; RMSEA = 0.075; CFI = 0.945; TLI = 0.936; SRMR = 0.046) (Hu & Bentler, 1999; Kline, 2010). Next, construct validity was also confirmed by assessing the convergent validity and discriminant validity. All CFA loadings were higher than 0.5, which provided evidence for convergent validity (Kline, 2010), and the average variance extracted (AVE) for each construct was greater than 0.5 (Anderson & Gerbing, 1988), suggesting that each construct is well represented by its own indicators. In addition, discriminant validity between constructs was also established because all AVEs ranging from 0.619 to 0.883 exceeded squared correlations between the constructs, ranging from 0.001 to 0.490 (Fornell & Larcker, 1981). Cronbach's alpha estimates of all constructs ranged from 0.751 to 0.973, which exceeded the recommended 0.70 cutoff point (Cortina, 1993), confirming the reliability of measurement. Tables 2 and 3 present the results regarding convergent and discriminant validities and the reliability of the instruments.

Table 2 Measurement model results

Constructs/Indicators	CFA loading	Cronbach's α	AVE
Adventure shopping		0.751	0.622
Go out for shopping is always exciting for me	0.698		
"I am living in my own world" describes my feeling towards shopping	0.870		
Idea shopping		0.927	0.815
I can follow the current fashion trend through shopping	0.877		
I can learn new fashion trends though shopping	0.977		
I can get to know new products through shopping	0.850		
Achievement shopping		0.855	0.619
Get the products in the shopping list purchased is very important in a shopping experience	0.805		
Be able to buy what I planned to buy is important for me in a shopping experience	0.895		
I feel good when I bought what I planned to buy in a shopping experience	0.830		
I always like my shopping plan, it makes me feel I am productive and I am a smart consumer	0.582		
Perceived enjoyment		0.961	0.832
I feel that co-designing clothing can be very interesting	0.881		
I enjoy the process of co-designing clothing	0.883		
I believe that co-designing clothing is joyful	0.925		
It is a fun game to customize my own clothing	0.939		
Co-designing clothing is a delightful experience	0.931		
Attitude (I think that co-designing clothing is…)		0.973	0.883
Bad: Good	0.950		
Foolish: Wise	0.954		
Unpleasant: Pleasant	0.964		
Unsatisfying: Satisfying	0.951		
Unfavorable: Favorable	0.876		
Subjective norms		0.948	0.860
My family and friends think co-designing clothing is a good idea for me	0.906		
The people who are important to me think I should co-design the clothing I like	0.936		
Those people who are important to me will consider co-designing their own clothing	0.939		

(continued)

Table 2 (continued)

Constructs/Indicators	CFA loading	Cronbach's α	AVE
Perceived behavior control		0.952	0.819
I believe I have the ability to co-design my favorite clothing	0.908		
In the process of co-design, I am very clear what is correct	0.956		
I think I can manage co-designing my own clothing handily	0.902		
I think co-designing my own clothing is in my control	0.852		

Note CFA loading is standardized

Table 3 Discriminant and convergent validity of constructs

	1	2	3	4	5	6	7
1. Adventure shopping	0.622						
2. Idea shopping	0.002	0.815					
3. Achievement shopping	0.001	0.020	0.619				
4. Perceived enjoyment	0.092	0.280	0.067	0.832			
5. Attitude	0.074	0.289	0.073	0.178	0.883		
6. Subjective norms	0.490	0.193	0.005	0.007	0.003	0.860	
7. Perceived behavior control	0.020	0.432	0.108	0.113	0.130	0.011	0.819

Note The numbers below diagonal are the squared correlation coefficient between the variables. The numbers in diagonal are the average variance extracted by each variable

Structural Model and Hypotheses Test

A structural equation model (SEM) was then conducted to examine the influence of adventure shopping, achievement shopping, idea shopping, perceived enjoyment, attitude, subjective norms, and perceived behavioral control on Chinese consumers' intention to purchase customized apparel products. The statistic results revealed an acceptable model fit ($\chi^2_{(df=304)} = 837.877$, $p < 0.000$, $\chi^2/df = 2.76$; RMSEA = 0.078; CFI = 0.937; TLI = 0.927; SRMR = 0.054). The outputs demonstrated that idea shopping was positively associated with subjective norms ($\beta = 0.093$, $p < 0.040$), perceived behavioral control ($\beta = 0.137$, $p < 0.016$), and perceived enjoyment ($\beta = 0.151$, $p < 0.014$), but had no influence on attitude ($\beta = 0.008$, $p < 0.796$). Therefore, H2 was partly supported. Neither achievement shopping value nor adventure shopping value had an influence on any of the TPB variables. However, both adventure shopping ($\beta = 0.141$, $p < 0.015$) and achievement shopping ($\beta = 0.331$, $p < 0.000$) were found to be positively associated with perceived enjoyment. Thus, both H1 and H3 were partly supported. Furthermore,

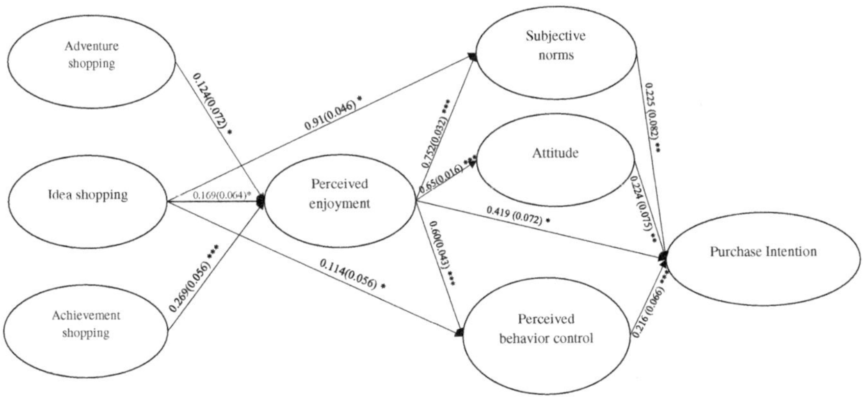

Fig. 1 Structural model test results. *Note* ***$p < 0.001$, **$p < 0.01$, *$p < 0.05$; only significant relationships are presented; standardized estimates with standard errors in parentheses

perceived enjoyment was positively related to attitude ($\beta = 0.636$, $p < 0.000$), subjective norms ($\beta = 0.778$, $p < 0.000$), and perceived behavioral control ($\beta = 0.651$, $p < 0.001$). Therefore, H4 was supported. A positive relationship between purchase intention toward apparel customization with perceived enjoyment ($\beta = 0.532$, $p < 0.016$) was also found among Chinese consumers. Thus, H5 was supported. In addition, the results exhibited significantly positive relationships between purchase intention toward apparel customization and attitude ($\beta = 0.268$, $p < 0.003$), subjective norms ($\beta = 0.275$, $p < 0.006$), perceived behavioral control ($\beta = 0.256$, $p < 0.001$). Therefore, H6 was supported. Figure 1 summarizes the hypotheses test results.

Discussion and Implications

Textile waste has been an increasing concern facing sustainability in China. Customization, an alternative retail model in the apparel industry, may help to increase product longevity because participating in the process will create a deeper emotional connection between the consumer and the product (Lee & DeLong, 2017). The emotional attachment may encourage people to keep the customized apparel products for a longer period of time (Armstrong et al., 2015). The purpose of this study is to determine the factors that influence Chinese consumers' intention to purchase customized apparel products. Three shopping values and perceived enjoyment were investigated in this study. Further, drawing upon the theory of planned behavior (TPB), the study developed a model to examine the influence of individual factors, including attitude and perceived behavioral control, as well as the social factor, such as subjective norm, on the intention to purchase customized apparel. The results point out the positive influence of shopping values and perceived enjoyment on Chinese consumers' purchase intention toward apparel customization.

Specifically, study results illustrate that perceived enjoyment of apparel customization is positively influenced by adventure shopping, idea shopping, and achievement shopping. Shopping values are critical psychological factors that encourage consumers to adopt a certain behavior (Solomon & Rabolt, 2006). Consumers are motivated not only by the utilitarian value but also for the emotional satisfaction obtained during the shopping process. People with adventure shopping value would like to seek freshness and excitement in order to escape from the aversive state of boredom (McGuire, 1974). Apparel customization requires consumers to participate in the co-design process, which may bring excitement to those consumers with adventure shopping value; therefore, they may perceive the process enjoyable. Similarly, individuals with idea shopping value tend to go shopping to keep up with new fashion trends, new products, and innovations (Arnold & Reynolds, 2003). Apparel customization requires that consumers be innovative and have creative ideas. In addition, the co-design process in customization requires consumers to communicate and interact with the professional designer or retailer, which may provide consumers with a chance to gain knowledge about new style and fashion ideas and fuel consumers' enjoyment of the process. Achievement shopping values reflect the importance of accomplishment in what had been planned on a particular shopping trip (Kim, 2006). Apparel customization provides consumers with an opportunity to get involved in the process of designing and making one-of-a-kind apparel products for themselves. Through apparel customization, consumers' specific need for a unique style or better fit can be achieved. A feeling of accomplishment may convey an excitement to those consumers with achievement shopping values. To encourage more apparel customization among Chinese consumers, enhancing the entertainment aspect of co-design and emphasizing the enjoyment and benefits of learning new fashion ideas would be beneficial for the development of customization businesses. In order to stimulate Chinese consumers to participate in apparel customization, marketers and retailers need to highlight that the process of customization is going to be an adventurous and exciting experience. By participating in the process, consumers will not only learn new fashion trends and style but also be able to get apparel products with better fit and unique style. Consumers will also feel a sense of achievement by learning new trends and being productive in the shopping journey. In addition, retailers also could highlight the benefit of gaining pride of authorship through apparel customization, which may increase the perceived enjoyment of customization process and attract those consumers who put a high value on the shopping achievement.

As predicted, the positive relationship between perceived enjoyment and intention to purchase customized apparel products was confirmed, indicating that the perceived enjoyment of customization or co-design process increases consumers' intention to participate in apparel customization, and they would be more likely to get involved in customizing their own apparel products. This result is consistent with previous research (Childers et al., 2001; Lee & Chang, 2011), confirming the positive influence of perceived enjoyment. In addition, the positive influences of perceived enjoyment on subjective norms, attitude, and perceived behavior control were also confirmed. Enjoyable feeling is an important motivator for people to get involved in a certain behavior. Individuals may engage in a particular behavior and be more confident about their behavior if it yields fun and enjoyment. Customization requires the participation of consumers in the process of co-design. In order to achieve the particular style they anticipate, consumers may need to try different choices provided by retailers. Not enjoying the activity of customization may discourage consumers' engagement in the process. Marketing for apparel customization could highlight the benefit of self-design and the rewards obtained from the co-design process, which might add value to apparel customization. In addition, retailers also can emphasize that necessary assistance from professional designers will be available during the customization process, which is fun, creative, and interesting.

As stated earlier, consumers' intention to purchase customized apparel products needs to be understood and explained by associating both personal shopping values and social factors. The theory of planned behavior (TPB), taking into account both individual and social factors, was adopted as the theoretical framework to analyze the determinants of Chinese consumers' intention toward purchasing customized apparel products. Drawing upon TPB, the intention to purchase customized apparel was proposed to be related to attitudes, subjective norms, and perceived behavioral control. As predicted, the results confirmed that the three factors in the TPB model contribute significantly to the prediction of consumers' intentions to purchase customized apparel products. The results suggest that the decision to get involved in apparel customization is a rational process wherein consumers take into account their attitudes and abilities. Getting involved in the process of apparel customization requires consumers to be creative and to be confident about their personal style and abilities. Some technology skills may also be needed if they customize apparel products online. Individuals who have more positive beliefs about their abilities and skills in the process of apparel customization would be more confident in their behaviors; therefore, they would be more likely to participate in apparel customization in the future. The positive influence of subjective norms also correlates with the collectivism core cultural value of Chinese consumers who tend to be interdependent and are conformists (Hofstede, 1980). Shaping attitudes and reducing perceived difficulties of performing apparel customization are crucial as they may help consumers develop a sense of personal confidence regarding apparel customization. For example, retailers can highlight that professional assistance from designers will help to make the co-design process easy and fun.

Unexpectedly, although both adventure shopping and achievement shopping have no influence on any of the TPB variables, the influence of idea shopping on subjective norms and perceived behavior control turned out to be significantly positive. This result indicates that individuals who tend to gather information about new trends through shopping usually have strong beliefs about their behaviors and are concerned with social pressure from others with regard to apparel customization. With the positive influence of subjective norms and perceived behavioral control on the purchase intention toward apparel customization, the indirect influence of idea shopping on the intention was confirmed. One suggestion for fashion retailers who provide customization services would be to offer videotape services by taking a video for consumers during the co-design process that captures a memorable moment. This may encourage more consumers to participate in apparel customization and enhance the entertainment aspect of co-design, as well as the value of learning new ideas. In addition, customization retailers also can emphasize the enjoyment and benefits of learning new fashion ideas in their marketing strategy, which will be beneficial for the promotion of apparel customization businesses.

Conclusion and Limitations

Apparel customization may present a solution to reduce excessive textile waste by increasing product longevity through building a solid emotional bond between the consumer and customized products (Medini et al., 2015), because it results in personal attachment to customized styles and well-fitted garments (Armstrong et al., 2015), which might be kept for a longer period of time. Although customization has generated much interest in the fashion industry, not many Chinese consumers get involved in the process of apparel customization. This research adds a unique contribution to the body of knowledge about Chinese consumers' purchase intention toward apparel customization by identifying the positive role of three shopping values in the process of customization. The results also highlight the positive influence of perceived enjoyment, attitude, perceived ability, and subjective norms on the intention to purchase customized apparel products, as well as on the relationships between shopping values and the intention.

In addition, the study successfully extended the TPB model by taking shopping values and perceived enjoyment as the external factor in this model. Through a structural equation estimation, idea shopping value and perceived enjoyment were proved to play an important role in the development of consumer attitude and subjective norm, thereby influencing Chinese consumers' intention to purchase customized apparel products.

Moreover, apparel customization serves to increase the longevity of the consumer–product relationship. It encourages consumers to be more sustainable by increasing product longevity without requiring a pro-environmental attitude, which may also provide marketers and retailers with new directions in which to support sustainability and simultaneously offer a new and innovative business model for meeting fashion needs while balancing the environmental costs.

This study has several limitations, and further study may be needed in future. Firstly, a convenience sample and online survey method were used to collect data in China, which might limit the generalizability of the results. Second, the retail concept of apparel customization was explained in one sentence; participants may have different interpretations as they pertain to understanding apparel customization, and this interpretation may have an influence on their response to the willingness to purchase customized products and generate bias. Third, although behavioral intention has been used to measure behavior, the results are based on consumers' intention, rather than behavior. Longitudinal research, which traces the model in association with behavior, would complement the findings of this study.

References

Abnett, K. (2015). *Will mass customization works for fashion*? Retrieved from https://www.businessoffashion.com/articles/intelligence/mass-customisation-fashion-nike-converse-burberry.

Ajzen, I. (1991). The theory of planned behavior. *Organizational Behavior and Human Decision Processes, 50*(2), 179–211.

Ajzen, I. (2002). *Constructing a TpB questionnaire: Conceptual and methodological considerations*. Retrieved from http://www.uni-bielefeld.de/ikg/zick/ajzen%20construction%20a%20tpb%20questionnaire.pdf.

Ajzen, I., & Fishbein, M. (1980). *Understanding attitudes and predicting social behavior*. Englewood Cliffs, NJ: Prentice-Hall.

Anderson, J. C., & Gerbing, D. W. (1988). Structural equation modeling in practice: A review and recommended two-step approach. *Psychological Bulletin, 103*(3), 411–423.

Armstrong, C. M., Niinimäki, K., Kujala, S., Karell, E., & Lang, C. (2015). Sustainable fashion product service systems: An exploration in consumer acceptance of new consumption models. *Journal of Cleaner Production, 97*(15), 30–39.

Arnold, M. J., & Reynolds, K. E. (2003). Hedonic shopping motivations. *Journal of Retailing, 79*(2), 77–95.

Atsmon, Y., Magni, M., Li, L., & Liao, W. (2012). *Meet the 2020 Chinese consumer*. Retrieved from http://www.mckinsey.com/global-themes/asia-pacific/meet-the-chinese-consumer-of-2020.

Birtwistle, G., & Moore, C. M. (2007). Fashion clothing-where does it all end up? *International Journal of Retail & Distribution Management, 35*(3), 210–216.

Boer, C. R., Pedrazzoli, P., Bettoni, A., & Sorlini, M. (2013). *Mass customization and sustainability: An assessment of framework and industrial implementation*. London, UK: Springer.

Bruno, T. D., Nielsen, K., Taps, S. B., & Jorgensen, K. A. (2013). Sustainability evaluation of mass customization. *International Federation for Information Processing, 414,* 175–182.

Chapman, J. (2009). Design for (emotional) durability. *Design Issues, 25*(4), 29–35.

Childers, T. L., Carr, C. L., Peck, J., & Carson, S. (2001). Hedonic and utilitarian motivations for online retail shopping behavior. *Journal of Retailing, 77*(4), 511–535.

Cho, H., & Fiorito, S. S. (2009). Acceptance of online customization for apparel shopping. *International Journal of Retail & Distribution Management, 37*(5), 389–407.

Chung, Y. S. (2015). Hedonic and utilitarian shopping values in airport shopping behavior. *Journal of Air Transport Management, 49,* 28–34.

CIW Team. (December 29, 2016). *Top WeChat stats and trends 2016*. Retrieved from https://www.chinainternetwatch.com/19524/wechat-data-report-2016/.

Cortina, J. M. (1993). What is coefficient alpha? An examination of theory and applications. *Journal of Applied Psychology, 78*(1), 98–104.

Davis, F. D., Bagozzi, R. P., & Warshaw, P. R. (1992). Extrinsic and intrinsic motivation to use computers in the workplace. *Journal of Applied Social Psychology, 22*(14), 1111–1132.

Fiore, A. M., Lee, S., & Kunz, G. (2003). Psychographic variables affecting willingness to use body-scanning. *Journal of Business and Management, 9*(3), 271–287.

Fiore, A. M., Lee, S.-E., & Kunz, G. (2004). Individual differences, motivations, and willingness to use a mass customization option for fashion products. *European Journal of Marketing, 38*(7), 835–849.

Flynn, A., & Vencat, E. F. (2012). *Custom nation: Why customization is the future of business and how to profit from it.* Dallas, TX: BenBella Books.

Fornell, C., & Larcker, D. F. (1981). Evaluating structural equation models with unobservable variables and measurement error. *Journal of Marketing Research, 18*(1), 39–50.

Franke, N., & Schreier, M. (2008). Product uniqueness as a driver of customer utility in mass customization. *Market Lett, 19,* 93–107.

Franke, N., & Schreier, M. (2010). Why customers value self-designed products: The importance of process effort and enjoyment. *Journal of Product Innovation Management, 27*(7), 1020–1031.

Hofstede, G. (1980). *Culture's consequences.* Beverly Hills, CA: Sage.

Hu, L.-T., & Bentler, P. M. (1999). Cutoff criteria for fit indexes in covariance structure analysis: Conventional criteria versus new alternatives. *Structural Equation Modeling, 6*(1), 1–31.

Irani, N., & Hanzaee, K. H. (2011). The effects of variety-seeking buying tendency and price sensitivity on utilitarian and hedonic value in apparel shopping satisfaction. *International Journal of Marketing Studies, 3*(3), 89–103.

Jackson, D. L., Gillaspy, J. A. J., & Purc-Stephenson, R. (2009). Reporting practices in confirmatory factor analysis: An overview and some recommendations. *Psychological Methods, 14*(1), 6–23.

Jayawardhena, C. (2004). Personal values' influence on e-shopping attitude and behavior. *Internet Research, 14*(2), 127–138.

Kamali, N., & Loker, S. (2002). Mass customization: On-line consumer involvement in product design. *Journal of Computer-Mediated Communication, 7*(4). Retrieved from http://onlinelibrary.wiley.com.fscproxy.framingham.edu/doi/10.1111/j.1083.

Kawamura, Y. (2005). *Fashionology: An introduction to fashion studies.* Oxford, England: Berg.

Kesari, B., & Atulkar, S. (2016). Satisfaction of mall shoppers: A study on perceived utilitarian and hedonic shopping values. *Journal of Retailing and Consumer Services, 31,* 22–31.

Kidwell, B., & Jewell, R. D. (2003). An examination of perceived behavioral control: Internal and external influences on intention. *Psychology & Marketing, 20*(7), 625–642.

Kim, H.-S. (2006). Using hedonic and utilitarian shopping motivations to profile inner city consumers. *Journal of Shopping Center Research, 13*(1), 57–79.

Kim, H.-S., & Hong, H. (2011). Fashion leadership and hedonic shopping motivations of female consumers. *Clothing and Textiles Research Journal, 29*(4), 314–330.

Kline, R. B. (2010). *Principles and practice of structural equation modeling* (3rd ed.). New York, NY: The Guilford Press.

Lau, H. F., Sin, L. Y. M., & Chan, K. K. C. (2005). Chinese cross-border shopping: An empirical study. *Journal of Hospitality & Tourism Research, 29*(1), 110–133.

Lawless, E., & Medvedev, K. (2016). Assessment of sustainable design practices in the fashion industry: Experiences of eight small sustainable design companies in the Northeastern and Southeastern United States. *International Journal of Fashion Design, Technology and Education, 9*(1), 41–50.

Lee, H.-H., & Chang, E. (2011). Consumer attitudes toward online mass customization: An application of extended technology acceptance model. *Journal of Computer-Mediated Communication, 16,* 171–200.

Lee, Y. K., & DeLong, M. (2017). American and Korean youths' attachment to handcraft apparel and its relation to sustainability. *Clothing and Textiles Research Journal, 35*(2), 67–80.

McGuire, W. J. (1974). Psychological motives and communication gratification. In J. G. Blumler & E. Katz (Eds.), *The uses of mass communications: Current perspectives on gratifications research.* Beverly Hills, CA and London: Sage.

McKinney, E., Gill, S., Dorie, A., & Roth, S. (2017). Body-to-pattern relationships in women's trouser drafting methods: Implications for apparel mass customization. *Clothing and Textiles Research Journal, 35*(1), 16–32.

Medini, K., Da Cunha, C., & Bernard, A. (2015). Tailoring performance evaluation to specific industrial contexts—Application to sustainable mass customization enterprises. *International Journal of Production Research, 53*(8), 2439–2456.

Merle, A., Chandon, J.-L., Roux, E., & Alizon, F. (2010). Perceived value of the mass-customized product and mass customization experience for individual consumers. *Production and Operation Management, 19*(5), 503–514.
Niinimäki, K. (2014). Sustainable consumer satisfaction in the context of clothing. In C. Vezzoli, C. Kohtala, & A. Srinivasan (Eds.), *Product-service system design for sustainability*. Sheffield: Greenleaf.
Niinimäki, K., & Hassi, L. (2011). Emerging design strategies in sustainable production and consumption of textile and clothing. *Journal of Cleaner Production, 19,* 1876–1883.
Ou, Y. (2011). *Apparel mass customization: Perceptions of young Chinese consumers*. Master Thesis, the University of North Carolina at Greensboro.
Peterson, J. (2016). The co-design process in mass customization of complete garment knitted fashion products. *Journal of Textile Science & Engineering, 6*(4), 270–278.
Piller, F. T., & Müller, M. (2004). A new marketing approach to mass customization. *International Journal of Computer Integrated Manufacturing, 17*(7), 583–593.
Piller, F. T., Schubert, P., Koch, M., & Moslein, K. (2005). Overcoming mass confusion: Collaborative customer co-design in online communities. *Journal of Computer-Mediated Communication, 10*(4). https://doi.org/10.1111/j.1083-6101.2005.tb00271.x.
Piller, F. T., & Tseng, M. M. (2010). *Handbook of research in mass customization and personalization*. Singapore: World Scientific.
Pine, B. J. I. I. I. (1993). *Mass customization: The new frontier in business competition*. Boston, MA: Harvard Business School Press.
Ravenhall, L. (2012). *Fast fashion and the destruction of developing countries*. Retrieved from http://www.ecooutfitters.co.uk/blog/fast-fashion-the-destruction-of-developing-countries/.
Salazar, H. A., Oerlemans, L., & van Stroe-Biezen, S. (2013). Social influence on sustainable consumption: Evidence from a behavioral experiment. *International Journal of Consumer Studies, 37*(2), 172–180.
Salvador, F., De Holan, P. M., & Piller, F. (2009). Cracking the code of mass customization. *MIT Sloan Management Review, 50*(3), 71–78.
Sanders, E. B. N., & Stappers, P. J. (2015). Probes, toolkits and prototypes: Three approaches to making in codesigning. *CoDesign: International Journal of CoCreation in Design and the Arts, 10*(1), 5–14.
Schreier, M. (2006). The value increment of mass-customized products: An empirical assessment. *Journal of Consumer Behavior, 5,* 317–327.
Senanayake, M. M., & Little, T. J. (2010). Mass customization: Points and extent of apparel customization. *Journal of Fashion Marketing and Management, 14*(2), 282–299.
Sherman, L. (2014). *Customizable fashion is on the rise*. Retrieved from http://fashionista.com/2014/06/customization-apparel-fashion.
Solomon, M. R., & Rabolt, N. J. (2006). *Consumer behavior in fashion*. Upper Saddle River, NJ: Prentice-Hall.
Steen, M., Manschot, M., & Koning, N. D. (2011). Benefits of co-design in service design project. *International Journal of Design, 5*(2), 53–60.
Teo, T. S. H., Lim, V. K. G., & Lai, R. Y. C. (1999). Intrinsic and extrinsic motivation in internet usage. *International Journal of Management Science, 27,* 25–37.
To, P. L., Liao, C., & Lin, T. H. (2007). Shopping motivations on Internet: A study based on utilitarian and hedonic value. *Technovation, 27*(12), 774–787.
Tseng, M. M., & Piller, F. (2003). *The customer centric enterprise: Advances in mass customization and personalization*. New York, NY: Springer.
Xu, Y., Chen, Y., Burman, R., & Zhao, H. (2014). Second-hand clothing consumption: A cross-cultural comparison between American and Chinese young consumers. *International Journal of Consumer Studies, 38*(6), 670–677.
Xu, Y., & McGehee, N. G. (2012). Shopping behavior of Chinese tourists visiting the United States: Letting the shoppers do the talking. *Tourism Management, 33*(2), 427–430.
Xue, H. Y., Zhang, X. J., & Wang, Y. Q. (2014). Research on the disposal strategy of waste textiles. *Applied Mechanics and Materials, 522,* 817–820.

What Will Trigger a Non-buyer to Become a Buyer in China's Luxury Goods Market? Cultural and Demographic Influences

Dong Shen, Jingxi Qian and Ying Jiang

Abstract The purpose of this study is to identify what triggers Chinese consumers to become a luxury goods buyer from a non-buyer. The objectives of this study are: (1) to compare Chinese luxury goods buyers and non-buyers by examining their key differences; (2) to identify potential cultural and demographic determinants of Chinese consumers' final purchase of luxury goods; and (3) to explore the effective strategies of how to trigger more consumers to purchase luxury goods in China. A survey was conducted in summer 2016 in China through WeChat, a social media app, and a total of 1,549 returned questionnaires were useful. Factor analysis, MANOVA, and logistic regression were chosen for data analyses. The results show that Chinese luxury goods buyers and non-buyers are significantly different in both cultural orientation and demographics. Chinese luxury goods buyers are short-term oriented, individualism focused, and power/status focused, whereas non-buyers are more long-term oriented, collectivism focused, and less power/status focused. They show significant differences in education level, family income, and location of residence. Chinese luxury goods buyers are more likely to be consumers who have higher family income, have higher education level, and live in more advanced and developed cities. Non-buyers have strong interest in luxury goods and show strong purchase intention. The high price point is the main issue. When they are ready to make the first purchase, they are more likely to buy luxury watches, clothing,

D. Shen (✉)
Fashion Merchandising and Design, Department of Family and Consumer Sciences, College of Social Science and Interdisciplinary Studies, California State University, MS 6053, 6000 J Street, Sacramento, CA 95819, USA
e-mail: dshen@csus.edu

J. Qian
School of Fashion Communication, Beijing Institute of Fashion Technology, Beijing, People's Republic of China
e-mail: 506122856@qq.com

Y. Jiang
School of Business, Beijing Institute of Fashion Technology, Beijing, People's Republic of China
e-mail: fzyjy@bift.edu.cn

Y. Xu et al. (eds.), *Chinese Consumers and the Fashion Market*, Springer Series in Fashion Business, https://doi.org/10.1007/978-981-10-8429-4_2

cosmetics, or accessories in mainland China off-line. Implications and limitations are addressed.

Keywords Consumers · China · Luxury · Culture · Demographics

Introduction

Luxury goods is defined in different ways, such as a philosophical–sociological approach, a micro-economic approach, or a managerial approach. Wiedmann, Hennigs, and Siebels (2007) define luxury as a multidimensional construct, including value perception of financial dimension (price value), functional dimension (usability, quality, uniqueness), individual dimension (self-identity, hedonic value, materialistic value), and social dimension (conspicuousness, prestige). Heine and Phan (2011) address the current lack of consumer-oriented approach and develop a comprehensive definition of luxury goods—"luxury goods have more than necessary and ordinary characteristics compared to other products of their category, which include their relatively high level of price, quality, aesthetics, rarity, extraordinarity, and symbolic meaning."

After slowing sales due to factors such as fears of terrorist attacks and currency fluctuations that kept many tourists away from cities including Paris, the luxury goods market will return to growth with an estimate of a global personal luxury goods market of close to $300 billion in 2017 (Bain & Company, 2017). Being a relatively new comer to the world luxury goods market at the beginning of the twenty-first century, China has taken a more leading role recently. In 2012, China officially overtook the USA as the world's biggest consumer nation of luxury goods for the first time (Doran, 2013). China's luxury spending in 2015 was at $16.8 billion (Horton, 2016). Chinese consumers account for the largest portion of luxury goods purchases (31%), followed by Americans (24%) and Europeans (18%) (D'Arpizio, Levato, Zito, & Montgolfier, 2015). It is estimated that over the next five years, Chinese luxury consumers will continue to drive growth as they maintain the biggest bloc of luxury shoppers globally, and by 2020, Chinese shoppers will make 34% of the total sale (Flora, 2016).

Even though China has become a key player in the global luxury market, much attention in both academics and business practices has been paid to the Chinese consumers who have purchased luxury goods. The luxury goods buyers have been researched immensely on their purchase motivation, attitudes, purchase intention, determinants of their purchase, and profiling (Black, 2007; Prendergast & Wong, 2003; Roll, 2017; Wang, 2008). In contrast, little has been studied on the non-buyers. However, the ones who have not purchased luxury goods in China may have an enormous potential and they might become new buyers quickly. In 2009, 33% of the total luxury market in China came from existing customers, whereas 67% came from first-time buyers. In 2010, 37% of the sales came from existing customers compared to 63% came from new buyers (Bain Co., 2011). If about two out of three Chinese luxury goods consumers are first-time buyers, studying the non-buyers and identifying their potentials are critical and necessary. In 2013, about 2% of Chinese population bought 30% of all luxury goods worldwide (Said, 2013). Looking at China's market with close to 1.4 billion population, it is obvious that not everyone is or will be a luxury goods consumer. However, studying the 98% of 1.4 billion population to explore their potentials, especially to identify what kind of non-buyers in this group are more likely to become luxury goods buyers, will be tremendously valuable. Therefore, this study puts focus on the comparison between the buyers and non-buyers of luxury goods. Furthermore, this study goes beyond a simple comparison between the buyers and non-buyers by taking a step further with a purpose of identifying what can trigger the non-buyers to become luxury goods buyers. The objectives of this study are: (1) to compare Chinese luxury goods buyers and non-buyers and examine their key differences; (2) to identify potential cultural and demographic determinants of Chinese consumers' final purchase of luxury goods; and (3) to explore effective strategies of how to trigger more consumers to purchase luxury goods in China.

This project is unique and significant. First, by focusing on luxury goods non-buyers rather than buyers, this study can bring an important addition to the current literature by filling a missing component to the research on luxury market. This contribution to the academic world can not only add the missing part to the current literature, but enrich the research on luxury goods consumers through a more complete angle from both buyers and non-buyers. Second, the results of this project can be of great interest to luxury goods companies. Since China's luxury goods market has become a major target for many luxury brands, it is undeniably important to have a thorough understanding of that market, especially the potential market given the fact that two-thirds of the luxury goods buyers are first-time buyers in China. Specifically, this study focuses on Chinese consumers' cultural orientation and demographics, and examines whether the buyers and non-buyers show differences in these dimensions. Examining the key differences between the buyers and non-buyers and identifying the determinants which can trigger the non-buyers to become buyers will be significant contributions to the business practitioners.

Literature Review and Conceptual Framework

Purchase Intention Versus Actual Purchase

Among research on consumer behavior, purchase intention is often chosen rather than actual purchase mainly because it is a more feasible measurement. A similar pattern exists in the research on Chinese luxury goods consumers as well. Zhang and Kim (2013) examined factors that affected Chinese consumers' attitude toward purchasing luxury fashion goods and purchase intention. When a cross-cultural study was conducted by Bian and Forsythe (2012) to examine American and Chinese consumers, again, purchase intention was investigated rather than actual purchase. Li, Li, and Kambele (2012) tried a more realistic approach by examining Chinese consumers' willingness to pay for luxury fashion brands related to their fashion lifestyle and perceived value. But willingness to pay is still very different from the final actual purchasing behavior.

Compared to purchase intention, actual purchase behavior is more valuable and reliable. Even though research has approved that there is a correlation between purchase intention and final purchase (Bemmaor, 1995; Newberry, Kleinz, & Boshoff, 2003), these are still two different concepts. It is not unusual to see consumers who have a higher purchase intention end up not making the final purchase due to various reasons. It is also not surprising to see consumers who have a lower purchase intention eventually make the final purchase. Purchase intention is a strong, but imperfect, predictor of purchasing. The strength of this relationship can vary quite a bit. Sheppard, Hartwick, and Warshaw (1988) found that the average correlation for the intention–behavior relationship was 0.53 ranging from 0.15 to 0.92. Morwitz, Steckel, and Gupta (2007) took a further look at the factors which had impact on the correlation between purchase intention and actual purchase, and they found that intentions were more correlated with purchases for existing products than for new ones; and more correlated when consumers were asked to provide intentions to purchase specific brands than when they were asked to provide intentions to buy at the product category level. Even though luxury goods have been existing in the Western market for many decades, it is still relatively new in China. For many Chinese consumers, luxury brands, luxury products, and luxury lifestyle are still new concepts. In addition, this study was to examine the overall luxury goods market in China rather than a specific brand. Therefore, a relatively low correlation between purchase intention and actual purchase is unavoidable. Thus, by asking their actual purchase behavior rather than their purchase intention, this study identified whether a Chinese consumer was a luxury goods buyer or non-buyer by using two versions of questionnaires assigned accordingly. With two sets of data collected, comparing the buyers and non-buyers and investigating what can trigger the non-buyers to become buyers of luxury goods become possible and significant.

Cultural Influence

Culture has an important influence on human behavior, with the assumption that an individual's behavior usually reflects their cultural value system. Adler (1997) developed a cultural–behavior model showing a chain effect starting from cultural influence on human attitude, then from attitude to final behavior. It is inevitable that such a pervasive force will influence how individuals act in a consumer context, because consumers often see things through the lens of culture (Teimourpour & Hanzaee, 2011). One of the most recognized cultural frameworks is established by

Hofstede (2001), who introduced cultural framework based on five dimensions: power distance, individualism/collectivism, masculinity/femininity, uncertainty avoidance, and long-term orientation/short-term orientation. The framework provides scales from 0 to 100 for 76 countries for each dimension, and each country has a position on each scale. Although the country scores were originally produced in the early 1970s, over 200 external comparative studies and replications have proved that country ranking in his data is still valid (Hofstede, 2001). In addition, Hofstede's cultural dimensions not only have been validated by many studies for explaining national differences, but have been also used for examining cultural differences at the individual levels. His work has been cited over 85,000 times showing its popularity and impact (Tung & Verbeke, 2010).

According to Hofstede (2001), power distance index (PDI) refers to the acceptance degree of unequally distributed power in organizations or families. Power and inequality are extremely fundamental facts of any society. Even though all societies are unequal, some are more unequal than others. Cultures with a high PDI tend to be more accepting of social hierarchies and authority. The countries with the highest PDI are mostly Eastern countries, whereas the counties with the lowest PDI are mostly Western countries (Hofstede, 2001). In Hofstede's model (2001), China scored 80 out of 100 compared to the USA with a score of 40. This indicates that Chinese expect and accept a high degree of social inequality and authority, and in turn, they strive for higher social status and authority to gain and maintain more power (Lau, 1988). The social dimension of luxury goods puts specific emphases on conspicuousness and prestige (Jung & Shen, 2011). The association with the upper social class remains an important aspect of luxury today (Kapferer, 1997; Nueno & Quelch, 1998). Research has shown that in the new emerging sociological dynamics of China, status plays a much higher role than in more stable and mature societies (Khairullah & Khairullah, 2013). Therefore, Chinese consumers look for means to show off or boost their status. Among the super-rich, conspicuous consumption is common to show off their wealth; and among the middle-class, buying and wearing luxury goods is seen as a passport for entering elite social circles (Lin & Wang, 2010). Buying and consuming luxury goods is one of the effective tools because luxury goods symbolize high social class and prestigious social elite which are all associated with high income, high authority, and more power. Therefore,

H1a. There is a significant difference of PDI between the luxury goods buyers and non-buyers in China.

H1b. The higher PDI, the more likely for Chinese consumers to purchase luxury goods.

Collectivism refers to the degree of which individuals are integrated into groups and individualism (IDV) is on the opposite side (Hofstede, 2001). In collectivist societies where the group orientation is emphasized, individuals are likely to feel pressured to conform to group norms. On the individualist side, when the ties between individuals are loose, everybody is expected to focus on themselves—their own success, their own benefit, their own needs and wants, and their own goals.

People who can show their own uniqueness are often looked up to; thus, everyone wants to be unique including a unique appearance and wardrobe. Most Western societies such as the USA are known to be individualistic while most Eastern societies such as China are collectivistic as groups' objectives and success are more important than individual achievement. China only scores 20 out of 100 in Hofstede's cultural dimensions in contrast with the USA's score of 91 (Hofstede, 2001). Therefore, China is a society with a strong collectivism influence. Even today when Chinese younger generations are more influenced by the Western culture, their ways of expressing individualism remains understated within the context of the social barriers of a Confucian society. It is observed that because China is not a truly individual culture, even ambitious Chinese youth show off in an understated way so that the tone and the manner of these independent attitudes are subtle and slight (Keller, Wang, Cooney, Erstad, & Lu, 2015). Since luxury always has a high premium price which is not affordable by the vast majority of the population, this high price contributes to the exclusivity and rarity of the offer (Chen & Lamberti, 2015). Luxury goods are closely tied with uniqueness and individualism because the individual dimension of luxury goods focuses on self-identity, hedonic value, materialistic value, self-satisfaction, and self-accomplishment (Jung & Shen, 2011). Therefore,

H2a. There is significant difference of IDV between the luxury goods buyers and non-buyers in China.

H2b. The higher IDV, the more likely for Chinese consumers to buy luxury goods.

Masculinity (MAS) versus its opposite, femininity, refers to the distribution of roles between the genders (Hofstede, 2001). The women in feminine countries have the same modest, caring values as the men. In the masculine countries, they are somewhat assertive and competitive, but not as much as the men, so that these countries show a gap between men's values and women's values (Teimourpour & Hanzaee, 2011). The gender gap can be often reflected in many aspects of a society, such as employment opportunities, salary ranges, professional choices, and family roles. Another very visible aspect is through appearance in that women would dress more feminine whereas man would look more masculine. China's score in this dimension is 66 out of 100 compared to the USA with a score of 62 (Hofstede, 2001), which does not show a sizable gap, especially when compared to other dimensions (Shi & Wang, 2011). More importantly, luxury goods as a whole is neither masculine nor feminine. There are certain brands or certain product lines emphasizing masculinity and others emphasizing femininity. But no gender tendency exists in the luxury market as a whole. Because this study focuses on the luxury goods market as a whole rather than a particular luxury brand or product category, there should be no high correlation between MAS and luxury goods purchase. Therefore,

H3. There is no significant difference of MAS between the luxury goods buyers and non-buyers in China.

Uncertainty avoidance (UAI) deals with a society's tolerance for uncertainty and ambiguity (Hofstede, 2001). UAI is defined as the extent of feeling "threatened by uncertain or unknown situations" (Hofstede, 1980, p. 113) as members of a society may be more or less tolerant of uncertain and ambiguous than those of other societies. Unstructured situations are novel, unknown, surprising, and different from usual. Uncertainty-avoiding cultures try to minimize the possibility of such situations through strict laws and rules, safety and security measures, and any other means. The opposite type, uncertainty-accepting cultures try to have as few rules as possible and they are relativist and allow many currents to flow side by side (Teimourpour & Hanzaee, 2011). The Chinese culture scores 40 compared to the US score of 46 on this UAI dimension (Hofstede, 1980).

Luxury goods are new products to many Chinese, especially the non-buyers. However, what defines luxury goods is its highest-ranked quality. Buying luxury goods can warrant consumers the best quality, which can lower any purchasing uncertainty. Therefore,

H4a. There is a significant difference of UAI between the luxury goods buyers and non-buyers in China.
H4b. The higher UAI, the more likely for Chinese consumers to purchase luxury goods.

Long-term orientation (LTO) is the fifth dimension of Hofstede's framework. LTO stands for the fostering of virtues oriented toward future rewards, in particular perseverance and thrift. Its opposite pole, short-term orientation, stands for the fostering of virtues related to the present (Hofstede, 2001). Some characteristics of long-term orientation include persistence, leisure time not too important, save, and be thrifty, whereas short-term orientation emphasizes quick results, leisure time importance, and spending. China scored 118 in this dimension compared to the US score of 29 (Hofstede, 1980), showing that Chinese culture is extremely long-term orientated (Shi & Wang, 2011). Given the characteristics of luxury goods and luxury goods consumption, the extremely high price is definitely against the basic focus of long-term orientation on being thrifty. For consumers who live more for the present than for the future, big spending on luxury goods is more likely to happen than long-term oriented consumers. Therefore,

H5a. There is significant difference of LTO between the luxury goods buyers and non buyers in China.
H5b. The higher LTO, the less likely for Chinese consumers to buy luxury goods.

Demographics

The current literature has consistently shown consumer demographic impact on purchase behavior. Individual consumer's age, gender, income, education, marital status, location of residence, and other demographic variables can influence consumers on different levels and in different ways. Similarly, luxury goods purchase behavior can also be influenced by consumer demographics. When China's luxury goods market is studied, research has been mainly done by focusing on Chinese consumers who have bought luxury goods, such as Black (2007) and Wang (2008). Do non-buyers show significant differences in demographic variables when they are compared to the luxury goods buyers? Are there specific demographic variables that hold Chinese non-buyers back from making the final purchase of luxury goods? Here, the most common six demographic variables, income, education, age, gender, marital status, and location of residence, are discussed and examined.

Since luxury goods always have a high price point, consumers' income level is often associated with luxury goods purchase. In contrast with necessity goods, luxury goods are more expensive and are often bought by individuals who have a higher disposable income or greater accumulated wealth than the average (Heine & Phan, 2011). In China, the luxury goods buyers of the past were the ultra-rich (Ben-Shabat, 2015). The super-rich who have accumulated enormous amounts of wealth emerged in the last couple of decades created by the soaring Chinese economy. They could be coal mine owners from Shanxi Province or company owners from Wenzhou in Zhejiang Province, known for its tradition of running small businesses. They could also be real-estate developers who have made millions in big city property deals, or prominent factory owners from the Pearl River delta area, where many international brands have been manufacturing since the "open door" policy allowed them into the country (Anonymous, 2007). Therefore,

H6a. There is significant difference of income between the luxury goods buyers and non-buyers in China.

H6b. The higher the income, the more likely for Chinese consumers to buy luxury goods.

Education is often highly correlated with income level. Gregorio and Lee (2002) present empirical evidence on how education is related to income distribution in a panel data set covering a broad range of countries for the period between 1960 and 1990 by finding that a higher educational attainment plays a significant role in a higher income level. In China, education is seen as the most important tool by Chinese parents for their kids' future, because investing in education is considered to be an important strategy to promote higher standards of living in China (Heckman, 2003). Wang (2011) found that in China, education indeed reduces earnings inequality while increasing individuals' earnings. Also, Chinese who have higher education level tend to have more exposure to Western culture by traveling abroad, receiving Western education, accessing to Western social media, and through other channels. Therefore,

H7a. There is significant difference of education level between the luxury goods buyers and non-buyers in China.

H7b. The higher the education level, the more likely for Chinese consumers to buy luxury goods.

The global luxury market used to be dominated by relatively older consumers, and today they are still part of the picture (Ben-Shabat, 2015). In China, consumers aged 35–50 have formed their dominating role in the luxury market. Individuals aged 30–34 in China made the highest average gross income in 2015; however, the age segment 45–49 represented the largest proportion among Chinese in the top income segment (Roberts, 2016). According to the forecast, by 2030 the age group of 40–44 in China will become the most prominent among the country's top income earners, representing opportunities for luxury goods (Roberts, 2016).

Meanwhile, younger consumers, especially Millennials, have started to make up an increasingly large share of the luxury market in China. The popular nickname for this Millennials generation is the little emperors and empresses, referring to their status as the only child in their families. While older generations may be more price-conscious and less brand oriented, China's youngsters do not share the same views (Wassner, 2009). They are more likely to be compulsive in their spending due to the desire to seek power and prestige (Li, Jiang, An, Shen, & Jin, 2009). They tend to engage in conspicuous consumption such as purchasing luxury goods. Also, due to the family and competitive pressure, it is important for them to purchase products and brands to impress their peers (Zhao & Belk, 2007). Overall, China's younger generation has far more inclination to spend than to save. With great potentials in both older generation who have higher earnings and younger generation who has higher desire for luxury goods in China, the following hypothesis is developed:

H8. There is no significant age difference between the luxury goods buyers and non-buyers in China.

In the global luxury market, female consumers have outperformed male consumers (Stokburger-Sauer & Teichmann, 2013). However, rapid global growth in male apparel and grooming categories is helping to balance the scales (Ben-Shabat, 2015). In contrast, an opposite pattern is observed in China. Chinese luxury market has been mainly male driven because it has been the only market in the world where men consume many more luxury goods than women (Olivier, 2013). Men account for at least 55% of China's luxury market, well above the global average of 40% (O'Connell, 2013). However, recently, due to the impact of the government's anti-corruption measures, Chinese male consumers' dominating role in luxury spending due to their culture of sealing deal through gifting has changed. In 2013, female consumers accounted for around half of all Chinese luxury buyers, representing an enormous growth from 1995 when 90% of luxury buyers were male, and this trend has kept going (Dawson, 2014). Today, women in China are gaining economic independence and are reaching a point of having their own money to spend on luxury items. Each year, 76% of China's female college graduates aspire to management positions, compared with 52% of their US counterparts. And China is home to 11 of the world's 20 richest self-made women, and it boasts 153 female yuan billionaires (around $150 million) (Burkitt, 2011). In a market where male consumers used to be the more dominating buyers of luxury goods, and female consumers have gradually caught up in the recent years, the following hypothesis is developed:

H9. There is no significant gender difference between the luxury goods buyers and non-buyers in China.

Singles might have more disposable income to invest in themselves than in family and children. However, due to the traditional Chinese cultural influence, most Chinese get married before they turn 30. Single Chinese are often younger than married ones, which means that their annual incomes are lower as well. However, even though married Chinese are more likely to make more money, they have more responsibilities to their family and children, which may lead to less disposable income.

In addition, singles often pay more attention to their public image and are more self-focused, which makes them being more interested in luxury goods due to the social dimension and individual dimension of luxury goods. However, because married consumers in China tend to be older than singles and as a result, they are more likely to have a better-established career than singles, the work-setting pressure for successful and prestigious looks may push the married consumers to make purchases of luxury goods. Therefore, when both married Chinese and single Chinese have their desires for luxury goods, the following hypothesis is developed:

H10. There is no significant difference of marital status between the luxury goods buyers and non-buyers in China.

Location of residence can directly relate to consumer behavior due to a variety of reasons such as living standard, climate, regional culture. Given luxury goods characteristics such as premium price point, uniqueness, and rareness, Chinese consumers living in different parts of the country have shown visible differences in luxury goods consumption. Being the tier-one cities, Shanghai, Beijing, Guangzhou, and Shenzhen account for 30% of wealthy Chinese (McKinsey, 2011). Meanwhile, growing wealth beyond tier-one cities has been creating several geographic markets with sizable pools of luxury goods consumers. Many of the China's tier-two cities, such as Hangzhou and Nanjing, even smaller tier-two cities, such as Qingdao and Wuxi, have become large enough to justify the presence of luxury stores catering to them (Atsmon, Dixit, & Wu, 2011). The luxury goods market will remain concentrated in the top 36 major cities in China mainly due to their economic development, which will account for 74% of the market's growth and 76% of total luxury sales by 2015 (Atsmon et al., 2011). Therefore,

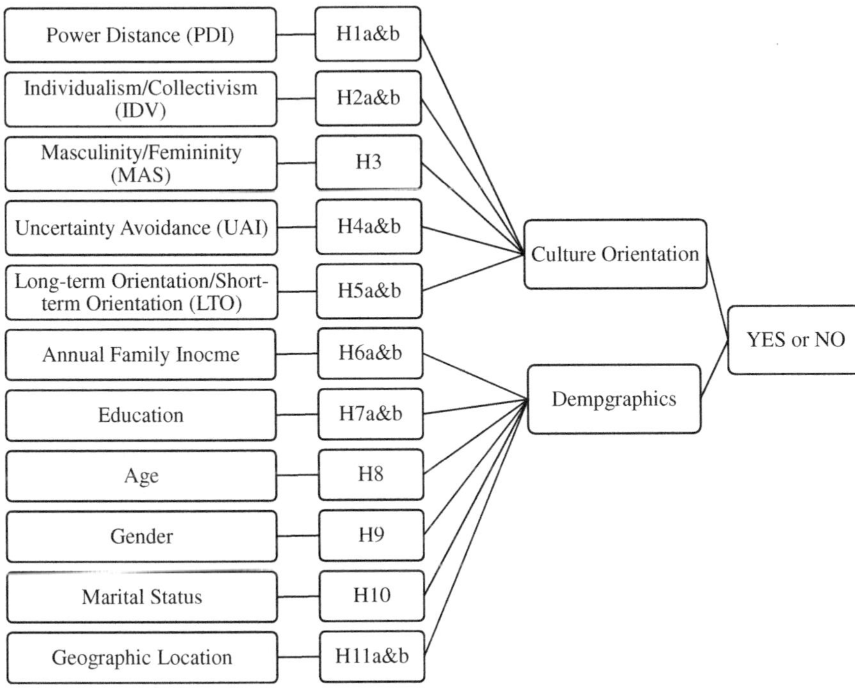

Fig. 1 Conceptual framework

H11a. There is significant difference of location of residence between the luxury goods buyers and non-buyers in China.

H11b. The more economic developed the location of residence, the more likely for Chinese consumers to buy luxury goods.

As a result, the following conceptual framework was developed and applied in this study (Fig. 1).

Methodology

A survey was conducted in the summer of 2016 for data collection. Two versions of questionnaires were developed. First, the participants were asked whether they had purchased luxury goods before. If their answer was "yes", they were led to Questionnaire 1; if their answer was "no", then they were assigned to Questionnaire 2. Questionnaire 1 consisted of three sections. The first section focused on the luxury goods purchase behavior such as luxury brands they have purchased and when they bought luxury goods for the first time. The second section measured the participants' cultural orientation by adopting CVSCALE. CVSCALE is a 26-item five-dimensional scale of individual cultural values that assesses Hofstede's cultural dimensions at the individual level, which has shown adequate reliability, validity, and across-sample and across-national generalizability (Yoo, Donthu, & Lenartowicz, 2011). "People in higher positions should make most decisions without consulting people in lower positions" and "People in higher positions should avoid social interaction with people in lower positions" are examples of PDI items. "Individuals should sacrifice self-interest for the group" and "Group success is more important than individual success" are examples of IDV items. "It is more important for men to have a professional career than it is for women" and "There are some jobs that a man can always do better than a woman" are examples of MAS items. "Standardized work procedures are helpful" and "Instructions for operations are important" are examples of UAI items. "Personal steadiness and stability" and "Giving up today's fun for success in the future" are examples of LTO items. The last section focused on demographics. In Questionnaire 2, the participants who have not purchased luxury goods before were measured on their possible behavior on luxury goods purchase, such as their willingness to purchase luxury goods, and the reasons why they have not made a purchase yet. Then the second section measured the participants' cultural orientation by using CVSCALE. The last section focused on demographics.

Both versions of questionnaire were originally written in English and then translated into Chinese by one of the researchers. Another two bilingual Chinese researchers confirmed the translation accuracy. The Chinese version of the questionnaire was then pre-tested by two groups of Chinese college students with 20 of each, one group for Questionnaire 1 and the other group for Questionnaire 2, for clarity and accuracy. Based on the feedback from the respondents, the questionnaires were slightly modified to make it clear and understandable. The answers from the pretest were not included in the main data analysis.

Because the main goal of this research is to study the potential luxury goods market in China, reaching out to Chinese consumers in various regions and places is critical. Therefore, an online survey became a better option. Specifically, WeChat, a social media app, was selected to distribute the questionnaire. WeChat, which is similar to Facebook, is a cross-platform instant messaging service developed in China in 2011. Being one of the largest stand-alone messaging apps, it has over a billion accounts and 700 million active users from all over China in 2016 (Wikipedia, 2016). College students and staff members at a university in Beijing were involved in the final data collection in the summer of 2016. On WeChat, each of them has their groups of friends, who reside all over the country in different cities and regions. They sent the survey link to each friend on WeChat and based on how the participant answered the first question, "whether you have purchased luxury goods before?", the app led the participant to the right version of the questionnaires. Total 55 students and staff were participants who sent the survey link to their WeChat groups, which generated a pool with potential 2,498 subjects. Eventually, 1,549 useful questionnaires were returned with a 62% return rate.

Results

Preliminary Results

As shown in Table 1, a total of 1,549 returned questionnaires were kept in the further data analyses. Among them, 656 (42.3%) participants were luxury goods buyers, whereas 893 (57.7%) were non-buyers. Combining both buyers and

Table 1 Demographic characteristics of the sample

	Buyers (656)		Non-buyers (893)	
	N	%	*N*	%
Family income				
<50,000 Yuan	66	10.1	238	26.7
50,000–100,000 Yuan	138	21.0	260	29.1
100,001–500,000 Yuan	266	40.5	324	36.3
500,001–1,000,000 Yuan	86	13.1	30	3.4
1,000,001–5,000,000 Yuan	61	9.3	19	2.1
>5,000,000 Yuan	39	5.9	22	2.5
Education				
Less than high school	13	2.0	39	4.4
High school graduates	50	7.6	119	13.3
College graduates	441	67.2	677	75.8
Graduate school and above	152	23.2	58	6.5
Age				
<20	15	2.3	34	3.8
20–30	299	45.6	452	50.6
31–40	178	27.1	138	15.5
41–50	75	11.4	142	15.9
51–60	74	11.3	92	10.3
>60	15	2.3	35	3.9
Gender				
Male	195	29.7	302	33.8
Female	461	70.3	591	66.2
Marital status				
Single	287	43.8	461	51.6
Married without children	67	10.2	62	6.9
Married with children	270	41.2	307	34.4
Divorced with children	12	1.8	21	2.4

(continued)

Table 1 (continued)

	Buyers (656)		Non-buyers (893)	
	N	%	*N*	%
Divorced without children	5	0.8	8	0.9
Others	15	2.3	34	3.8
Location				
First tier	306	46.6	318	35.6
Second tier	159	24.2	201	22.5
Third tier	146	22.3	260	29.1
Others	45	6.9	114	12.8

non-buyers, 624 (40.3%) participants live in tier-one cities, 360 (23.2%) live in tier-two cities, and 406 (26.2%) live in tier-three cities. In addition, 748 (48.3%) participants are single, and 577 (37.2%) are married with kids. Meanwhile, 1,052 (68.0%) are female, and 497 (32.0%) are male. The majority of the participants are aged 20-40 years and college graduates. In terms of family income, there is a quite big variety among the participants.

Reliability and Factor Analysis

Factor analysis was conducted with Varimax rotation to examine the cultural scale. With a five-factor structure, all five factors have an eigenvalue greater than 1 except for the fifth one, but its eigenvalue is 0.964, which is very close to 1. The total cumulative percentage of variance explained by all five factors combined is 75%, with a significant Bartlett's Test.

As shown in Table 2, six items are loaded on Factor 1, the long-term/short-term orientation dimension, with a Cronbach's Alpha of 0.933. Another six items are loaded on Factor 2, the individualism/collectivism dimension, with a Cronbach's Alpha of 0.941. Five items are loaded on Factor 3, power distance dimension, and the Cronbach's Alpha is 0.877. Another five items are loaded on Factor 4, the uncertainty avoidance dimension, and the Cronbach's Alpha is 0.926. Finally, the last four items are loaded on Factor 5, the masculinity/femininity dimension, with a Cronbach's Alpha of 0.871.

MANOVA

MANOVA was conducted to find out whether between the two groups, the buyers versus non-buyers, there were significant differences in their cultural orientation and demographics. The overall model was significant because the results showed that

Table 2 Results' summary of factor analysis and reliability check

	Number of items	Factor loading	% of variance	Cronbach's alpha
Factor 1 Long-term orientation/short-term orientation (LTO)	6	0.827	42.696	0.933
		0.818		
		0.814		
		0.735		
		0.731		
		0.702		
Factor 2 Individualism/collectivism (IDV)	6	0.834	14.812	0.941
		0.823		
		0.817		
		0.806		
		0.798		
		0.735		
Factor 3 Power distance (PDI)	5	0.828	8.975	0.877
		0.823		
		0.815		
		0.795		
		0.764		
Factor 4 Uncertainty avoidance (UAI)	5	0.785	4.850	0.926
		0.784		
		0.727		
		0.709		
		0.605		
Factor 5 Masculinity/femininity (MAS)	4	0.776	3.708	0.871
		0.775		
		0.730		
		0.730		

there was a statistically significant difference between the buyer and non-buyer groups, $F(11, 1537) = 24.764$, $p < 0.0005$; Wilk's $\Lambda = 0.849$, partial $\eta^2 = 0.151$.

Specifically, as shown in Table 3, there are significant differences between these two groups in long-term orientation/short-term orientation, individualism/collectivism, and power distance when cultural dimensions were studied. Therefore, H1a, H2a, H3, and H5a are all supported. However, H4a is not supported. When demographic impact was examined, the results showed that the buyers and non-buyers were significantly different in their education, income, and living location. H6a, H7a, and H11a are supported. Age, gender, and marital status did not show significant differences between the buyers and non-buyers. H8, H9, and H10 are supported.

Table 3 MANOVA results in comparison between buyers and non-buyers

Dependent variable	Type III sum of squares	*df*	Mean square	*F*	Sig.
Factor 1: Long-term/short-term orientation (LTO)	6.532	1	6.532	6.556	0.011*
Factor 2: Individualism/ collectivism (IDV)	9.888	1	9.888	9.946	0.002**
Factor 3: Power distance (PDI)	17.438	1	17.438	17.626	0.000***
Factor 4: Uncertainty avoidance (UAI)	2.305	1	2.305	2.307	0.129
Factor 5: Masculinity/ femininity (MAS)	0.980	1	0.980	0.980	0.322
Education	27.879	1	27.879	77.491	0.000***
Age	0.017	1	0.017	0.012	0.915
Gender	0.634	1	0.634	2.909	0.088
Marital status	1.920	1	1.920	1.266	0.261
Family income	217.276	1	217.276	153.583	0.000***
Residence of living	33.377	1	33.377	31.752	0.000***

$^{***}p < 0.001$. $^{**}p < 0.01$. $^{*}p < 0.05$

Logistic Regression

Even though MANOVA results showed that there were significant differences between the buyers and non-buyers in their cultural orientation and demographics, the relationship direction was still not clear. More important, what cultural orientation factors and demographic characteristics can be used as predictors to identify whether a Chinese consumer is a buyer or a non-buyer is of even greater interest and more valuable. Therefore, a logistic regression was conducted with all five cultural dimensions and six demographics as predictors to predict whether a Chinese consumer was a luxury product buyer or non-buyer.

The logistic regression model was statistically significant, $\chi^2(11, N = 1549) = 249.590$, $p = 0.000$. The Hosmer and Lemeshow test of the goodness of fit suggested the model was a good fit to the data as $p = 0.092$ (>0.05). The model explained 20% (Nagelkerke R^2) of the variance in DV and correctly classified 69.3% of the cases compared to 57.7% in the null model. Even though the Nagelkerke R^2 is relatively low, it is almost irrelevant because the main goal of this study is to determine which predictors are statistically significant and how changes in the predictors lead to the changes in the response variables. If any cultural factor or demographic factor has significant relationship with buying decision, this relationship would exist regardless of whether the R-squired value is 20% or 90%.

As shown in Table 4, the Wald test indicates that among the five cultural factors, long-term orientation/short-term orientation (LTO), individualism/collectivism (IDV), and power distance (PDI) are significant predictors. Specifically, the more

Table 4 Logistic regression results

Independent variable	B	S.E.	Wald	*df*	Sig.	Exp(B)
Factor 1: Long-term/short-term orientation (LTO)	0.177	0.056	9.929	1	0.002**	1.194
Factor 2: Individualism/ collectivism (IDV)	0.152	0.058	6.914	1	0.009**	1.164
Factor 3: Power distance (PDI)	−0.272	0.057	22.704	1	0.000***	0.762
Factor 4: Uncertainty avoidance (UAI)	0.035	0.056	0.388	1	0.533	1.036
Factor 5: Masculinity/ femininity (MAS)	0.014	0.059	0.057	1	0.812	1.014
Education	0.620	0.104	35.889	1	0.000***	1.860
Age	−0.026	0.057	0.199	1	0.656	0.975
Gender	0.363	0.126	2.346	1	0.104	0.838
Marital status	−0.084	0.054	2.441	1	0.118	0.920
Family income	0.474	0.050	89.381	1	0.000***	1.607
Residence of living	0.190	0.057	11.124	1	0.001**	1.209
Constant	−3.212	0.418	59.106	1	0.000	0.040

***$p < 0.001$. **$p < 0.01$

power distance a Chinese consumer pursues, the more likely of this consumer to make a luxury goods purchase. H1b is supported. The more individualism a Chinese consumer is, the more likely of this consumer to make a luxury goods purchase. H2b is supported. The more long-term oriented a Chinese consumer is, the less likely of this consumer to make a luxury goods purchase. H5b is supported. Again, H4a and H4b are not supported.

Among the six demographics, education, family income, and location are significant predictors. The higher family income a Chinese consumer has, the more likely of this consumer to make a purchase. H6b is supported. The higher education a Chinese consumer has, the more likely of this consumer to make a purchase. H7b is supported. The bigger and more developed a city a Chinese consumer lives in, the more likely of this consumer to make a purchase. H11b is supported.

Non-buyer Purchase Intention

The analyses above all show that there are significant differences between Chinese luxury goods buyers and non-buyers. By using effective predictors such as some cultural dimensions and demographic variables, it is possible to identify whether a Chinese consumer is a luxury goods buyer or a non-buyer. Being able to identify a buyer or non-buyer is not the ultimate goal though. What is more critical is to find out whether those non-buyers have interests in luxury goods and whether they have purchase intention to luxury goods.

The results show that among all the non-buyers, close to half of them (46.5%) chose "yes" when they were asked whether they would like to purchase luxury products, which shows strong purchase intention among the Chinese luxury goods non-buyers. Close to half of them (44.2%) would like to buy luxury goods as gifts for others and for themselves. Regarding product category, the non-buyers are most interested in buying luxury watches (46.4%), followed by clothing (44.7%), cosmetics (39.8%), and accessories (shoes, scarves, sunglasses, and belts) (37.4%). Regarding purchasing location, the non-buyers chose mainland China first (58.7%), followed by Hong Kong, Macau, and Taiwan (40.4%), and EU (26.4%) and USA (24.6%). And 43.1% of them chose to buy off-line rather than online.

Conclusions, Discussions, and Implications

The purpose of this study is to identify what triggers Chinese consumers to become a luxury goods buyer from a non-buyer. The objectives of this study are: (1) to compare Chinese luxury goods buyers and non-buyers by examining their key differences; (2) to identify potential cultural and demographic determinants of Chinese consumers' final purchase of luxury goods; and (3) to explore the effective strategies of how to trigger more consumers to purchase luxury goods in China. The results show that Chinese luxury goods buyers and non-buyers are significantly different in both cultural orientation and demographics. Specifically, they show significant differences in education level, family income, and location of residence. When cultural orientation is studied, they are different in long-term/short-term orientation, individualism/collectivism, and power distance. Chinese luxury goods buyers are often short-term oriented, individualism focused, and power/status focused, whereas non-buyers are more long-term oriented, collectivism focused, and less power/status focused. In addition, Chinese luxury products buyers are more likely to have higher family income, have higher education level, and live in more advanced and developed cities. Chinese luxury goods non-buyers have strong interest in luxury goods and show strong purchase intention as well. The high price point is the main reason why they have not made a purchase yet. When they are ready to make their first purchase, they are more likely to buy luxury watches, clothing, cosmetics, or accessories in mainland China off-line. All the hypotheses are supported by the data except for H4a and H4b on UAI. On one hand, it is true that the high price of luxury goods does provide a guarantee of high quality to consumers to avoid any uncertainty. On the other hand, however, without enough previous purchase experience and lacking the needed knowledge of luxury goods since luxury goods and brands are still relatively new to the China's market, especially for non-buyers, Chinese consumers who are high on UAI might be less likely to buy luxury goods to avoid uncertainty. With both possibilities mixed and combined, as a result, no significant difference of UAI was found between the luxury goods buyers and non-buyers in China.

Looking at China's market with close to 1.4 billion population, about 2% of Chinese population bought 30% of all luxury goods worldwide in 2013 (Said, 2013), then what about the rest 98%? In 2015, 35% of luxury goods sales came from Chinese consumers, followed by 15% from Americans and 10% from Japanese. It is estimated that in 2025, Chinese consumers will generate 50% of the sales (Rambourg, 2015). Therefore, a big portion of the 50% of the sales will come from first-time buyers. What set the luxury goods non-buyers apart from the buyers? Who are more likely to become the first-time buyers? This study shows that Chinese luxury goods non-buyers tend to have lower family income, have lower level of education, and live in less developed cities and regions in China. This is not a surprise because luxury goods are often associated with high premium price, high social status, and class. Therefore, further economic development and continuous increase of income, education, and living standard will be critical to trigger more non-buyers to become first-time buyers in China. Research has shown that in China, the share of urban households' annual income with $6,000–15,999 in 2000 was 63%, followed by a big increase to 82% in 2010. However, the share is estimated to drop to 36% in 2020 because 51% of the unban households will make annual income of $16,000–34,000, a level higher (Atsmon & Magni, 2012). This economic forecast provides the needed financial prerequisite to more Chinese consumers if they want to become luxury goods buyers. When more and more Chinese can earn higher income according to the forecast, with their strong willingness to purchase luxury goods, it is clear that many first-time buyers will emerge in China's luxury goods market in the next five years, which will eventually further expand China's luxury goods market greatly.

However, a higher level of income, a better education, and a higher level of living standard combined do not guarantee a non-buyer to become a buyer for sure. This study further proves that beyond the demographic differences, luxury goods non-buyers also differ from buyers in cultural orientations. Non-buyers are more long-term oriented, more collectivism focused, and less power/social status sensitive than luxury goods buyers. These findings clearly show the strategic directions to luxury goods companies in term of how to trigger more non-buyers to become buyers. Through advertising campaigns, selection of spokespersons, store concept, visual presentation, and media effects, luxury goods companies need to focus on searching for effective strategies to convert those long-term oriented, collectivism focused, and power/social status less sensitive non-buyers to the opposite. Because without these changes on cultural orientation, those non-buyers will not become buyers even if financially they can afford luxury goods. The two driving forces, cultural orientation and demographics, are both needed in order to trigger non-buyers to become buyers in China's luxury goods market.

Limitations and Future Studies

Despite key contributions research can bring, all studies have limitations including this one. First, an online survey through WeChat was conducted for data collection in order to reach out potential participants living in different regions and cities in China in a time-efficient and economic-efficient way. However, this survey strategy limits the age range of participants because older Chinese do not use WeChat as much as younger Chinese do. The result showed that more than 50% of returned questionnaires were completed by consumers between 20 and 30 years old. In 2016, the biggest two age-groups in China population are 20- to 30-year-old group and 40- to 50-year-old group (China 2016, 2016). The future studies may put more focus on the older group of Chinese consumers, especially the 40- to 50-year-old group, which are more career-established with a higher annual income compared to the younger groups.

Second, due to a small sample size (there are total 893 non-buyer questionnaires collected in this study), it is inappropriate to run a cluster analysis on the non-buyer group. However, segmenting and profiling the non-buyers will be invaluable for luxury goods businesses to market in China. Research (Black, 2007; Wang, 2008) has been done on categorizing Chinese luxury goods buyers for the development of different business strategies in order to target each segment more effectively. Similarly, it will be more beneficial and valuable if a cluster analysis can be conducted to segment all Chinese non-buyers into different groups. Future studies focusing on this direction will be needed.

References

Adler, N. J. (1997). *International dimensions of organizational behavior*. Cincinnati, OH: Shout-Western College Publishing.

Anonymous. (2007). Branding in China: Luxury looks east. *Brand Strategy, June 12*, 56.

Atsmon, Y., Dixit, V., & Wu, C. (April 2011). *Tapping China's luxury-goods market*. Retrieved from http://www.mckinsey.com/business-functions/marketing-and-sales/our-insights/tapping-chinas-luxury-goods-market.

Atsmon, Y., & Magni, M. (2012). *Meet the Chinese consumer of 2020*. Retrieved from http://www.mckinsey.com/global-themes/asia-pacific/meet-the-chinese-consumer-of-2020.

Bain & Company. (2017). *2017 Global luxury market report*. Retrieved from http://www.bain.com/about/press/press-releases/global-personal-luxury-goods-market-expected-to-grow-by-2-4-percent.aspx.

Bain Co. (2011). *2011 China luxury market study*. Retrieved from http://www.bain.com/Images/2011%20Bain%20China%20Luxury%20Market%20Study.pdf.

Bemmaor, A. C. (1995). Predicting behavior from intention-to-buy measures: The parametric case. *Journal of Marketing Research, 32*, 176–191.

Ben-Shabat, H. (2015). *The new luxury consumer? Think: Multiple consumers*. Retrieved from http://www.therobinreport.com/the-new-luxury-consumer-think-multiple-consumers/.

Bian, Q., & Forsythe, S. (2012). Purchase intention for luxury brands: A cross cultural comparison. *Journal of Business Research, 65*(10), 1443–1451.

Black, P. T. (2007). Luxury market in China is fast-growing and confusing. *Advertising Age, 78*(43), 46.

Burkitt, L. (2011). *In China, women begin splurging*. Retrieved from http://www.wsj.com/articles/SB10001424052748703864204576315291406475666.

Chen, S., & Lamberti, L. (2015). Entering the dragon's nest: Exploring Chinese upper-class consumers' perception of luxury. *Qualitative Market Research, 18*(1), 4–29.

China 2016. (2016). *Population pyramids of the word from 1950-2100*. Retrieved from https://populationpyramid.net/china/2016/.

Dawson, K. C. (2014). *Women are now taking over luxury shopping in China*. Retrieved from http://usa.chinadaily.com.cn/opinion/2014-02/04/content_17268466.htm.

D'Arpizio, C., Levato, F., Zito, D., & Montgolfier, J. (2015). *Luxury goods worldwide market study*. Retrieved from http://www.bain.com/Images/BAIN_REPORT_Global_Luxury_2015.pdf.

Doran, S. (2013). *2013 luxury industry predictions from the experts*. Retrieved from http://luxurysociety.com/articles/2013/01/2013-luxury-industry-predictions-from-the-experts.

Flora, L. (2016). *Bain: China's luxury market reverses decline with 2% growth*. Retrieved from https://jingdaily.com/bain-chinas-luxury-market-reverses-decline-2-growth/.

Gregorio, J. D., & Lee, J. (2002). Education and income inequality: New evidence from cross-country data. *The Review of Income and Wealth, 48*(3), 395–416.

Heckman, J. J. (2003). China's investment in human capital. *Economic Development and Cultural Change, 51,* 795–804.

Heine, K., & Phan, M. (2011). Trading-up mass-market goods to luxury products. *Australasian Marketing Journal, 19*(2), 108–114.

Hofstede, G. (1980). *Culture's consequences: International differences in work-related values*. Beverly Hills, CA: Sage Publications.

Hofstede, G. (2001). *Culture's consequences*. Thousand Oaks, CA: Sage Publications.

Horton, C. (2016). *When it comes to luxury, China still leads*. Retrieved from http://www.nytimes.com/2016/04/05/fashion/china-luxury-goods-retail.html?_r=0.

Jung, J., & Shen, D. (2011). Brand equity of luxury fashion brands among Chinese and U.S. young female consumers. *Journal of East-West Business, 17*(1), 48–69.

Kapferer, J. N. (1997). Managing luxury brands. *Journal of Brand Management, 4*(4), 251–260.

Keller, J., Wang, Y., Cooney, J., Erstad, A., & Lu, C. (2015). Cultural dimensions: A comparative analysis of aviation students in China and the U.S. *International Journal of Aviation, Aeronautics, and Aerospace, 2*(3). Retrieved from http://commons.erau.edu/ijaaa/vol2/iss3/3.

Khairullah, D. H. Z., & Khairullah, Z. Y. (2013). Cultural values and decision-making in China. *International Journal of Business, Humanities and Technology, 3*(2), 1–12.

Lau, S. (1988). The value orientations of Chinese university students in Hong Kong. *International Journal of Psychology, 23*(5), 583–596.

Li, D., Jiang, Y., An, S., Shen, Z., & Jin, W. (2009). The influence of money attitudes on young consumers' compulsive buying. *Young Consumers, 10*(2), 98–109.

Li, G., Li, G., & Kambele, Z. (2012). Luxury fashion brand consumers in China: Perceived value, fashion lifestyle, and willingness to pay. *Journal of Business Research, 65*(10), 1516–1522.

Lin, X., & Wang, C. L. (2010). The heterogeneity of Chinese consumer values: A dual structure Explanation. *Cross Cultural Management: An International Journal, 17*(3), 244–256.

Morwitz, V. G., Steckel, J., & Gupta, A. (2007). When do purchase intentions predict sales? *International Journal of Forecasting, 23*(3), 347–364.

McKinsey. (2011). *From 'Made in China' to 'Sold in China', the Rise of the Chinese Urban Consumer*. McKinsey Global Institute, McKinsey & Company.

Nueno, J. L., & Quelch, J. A. (1998). The mass marketing of luxury. *Business Horizons, 41*(6), 61–68.

Newberry, C. R., Kleinz, B. R., & Boshoff, C. (2003). Managerial implications of predicting purchase behavior from purchase intentions: A retail patronage case study. *Journal of Services Marketing, 17,* 609–618.

Olivier. (2013). *Chinese men buy more luxury goods than Chinese women*. Retrieved from http://marketingtochina.com/chinese-men-buy-more-luxury-goods-than-chinese-women/.

O'Connell, S. (2013). *Why men are so important to China's luxury market*? Retrieved from http://luxurysociety.com/articles/2013/02/why-men-are-so-important-to-chinas-luxury-market.

Prendergast, G., & Wong, C. (2003). Parental influence on the purchase of luxury brands of infant apparel: An exploratory study in Hong Kong. *The Journal of Consumer Marketing, 20*(2/3), 157–169.

Rambourg, E. (2015). *Meet the Chinese luxury shoppers who are taking over the world*. Retrieved from http://www.businessinsider.com/chinese-luxury-shoppers-are-taking-over-2015-1.

Roberts, E. (2016). *The latest facts and figures about the Chinese luxury market*. Retrieved from http://www.luxurysociety.com/en/articles/2016/08/the-latest-facts-and-figures-about-the-chinese-luxury-market/.

Roll, M. (2017). *Chinese luxury consumers—Trends and challenges for luxury brands*. Retrieved from https://martinroll.com/resources/articles/asia/chinese-luxury-consumers-trends-and-challenges-for-luxury-brands/.

Said, S. (2013). *2% of Chinese population buy 30% of all luxury goods*. Retrieved from http://www.therichest.com/business/economy/2-of-chinese-buy-30-of-all-luxury-goods/.

Sheppard, B. H., Hartwick, J., & Warshaw, P. R. (1988). The theory of reasoned action: A meta-analysis of past research with recommendations for modifications and future research. *Journal of Consumer Research, 15*, 325–343.

Shi, X., & Wang, J. (2011). Cultural distance between China and US across GLOBE Model and Hofstede Model. *International Business and Management, 2*(1), 11–17.

Stokburger-Sauer, N. E., & Teichmann, K. (2013). Is luxury just a female thing? The role of gender in luxury brand consumption. *Journal of Business Research, 66*(7), 889–896.

Teimourpour, B., & Hanzaee, K. H. (2011). The impact of culture on luxury consumption behaviour among Iranian consumers. *Journal of Islamic Marketing, 2*(3), 309–328.

Tung, R. L., & Verbeke, A. (2010). Beyond Hofstede and GLOBE: Improving the quality of cross-cultural research. *Journal of International Business Studies, 41*(8), 1259–1274.

Wassner, B. (2009). Young buyers challenge luxury retailers in Asia. *New York Times*, November 17.

Wang, J. (2008). Viewpoint: The language of chopsticks. *Journal of Advertising Research, 48*(4), 481–484.

Wang, L. (2011). *How does education affect the earnings distribution in urban China*? Retrieved from http://anon-ftp.iza.org/dp6173.pdf

Wiedmann, K. P., Hennigs, N., & Siebels, A. (2007). Measuring consumers' luxury value perception: A cross-cultural framework. *Academy of Marketing Science Review, 11*, 1–21.

Wikipedia. (2016). *WeChat*. Retrieved from https://en.wikipedia.org/wiki/WeChat.

Yoo, B., Donthu, N., & Lenartowicz, T. (2011). Measuring Hofstede's five dimensions of cultural values at the individual level: Development and Validation of CVSCALE. *Journal of International Consumer Marketing, 23*(3–4), 193–210.

Zhang, B., & Kim, J. (2013). Luxury fashion consumption in China: Factors affecting attitude and purchase intent. *Journal of Retailing and Consumer Services, 20*, 68–79.

Zhao, X., & Belk, R. W. (2007). Live from shopping malls: Blogs and Chinese consumer desire. *Advances in Consumer Research, 34*, 131–137.

Part II
Digital Retailing in China

How Web site Quality Affects Apparel Mobile Commerce Consumer Satisfaction and Intent to Purchase? A Study of Chinese Consumers

Ting Chi and Quinn Sullivan

Abstract In recent years, apparel mobile commerce has experienced a rapid development in China, which is largely fueled by proliferation of smartphones, decreased cost of mobile data, increasing number of m-tailers, and shift of consumer shopping style toward mobility and time flexibility. Some apparel m-tailers can attract millions of visitors to their web sites everyday while others seem not appealing to consumers. Thus, this study aims to identify the effects of key features of mobile web site quality on the Chinese consumer satisfaction that in turn affects their intent to shop apparel via mobile web sites. Six dimensions of mobile web site quality, including web site visual appeal, apparel visual appeal, brand trust, web site informativeness, web site response time, and web site security, were investigated. Using the primary data collected via an online survey, multiple regression analysis was applied to test the proposed relationships. Results reveal that web site informativeness, web site visual appeal, apparel visual appeal, and web site security positively affect consumer satisfaction toward apparel m-commerce web site while web site response time and brand trust show insignificant impacts on consumer satisfaction. With higher satisfaction on apparel m-commerce web site, consumers are more likely to purchase apparel through the web site.

Keywords M-commerce · Web site quality · Purchase intention

T. Chi (✉) · Q. Sullivan
Department of Apparel, Merchandising, Design & Textiles,
Washington State University, Pullman WA 99164, USA
e-mail: tchi@wsu.edu

Q. Sullivan
e-mail: quinn.sullivan@wsu.edu

Y. Xu et al. (eds.), *Chinese Consumers and the Fashion Market*, Springer Series in Fashion Business, https://doi.org/10.1007/978-981-10-8429-4_3

Introduction

China is the world's largest smartphone market, accounting for approximately 30% of the global smartphone market. Chinese smartphone users are expected to jump from 527 million in 2015 to 640 million in 2018, according to eMarketer, (2016). It has been witnessed that Chinese consumers are quickly getting used to use their gadgets to snap up a wide array of products and services via the Internet. In 2016, nearly half of all China's e-commerce sales, totaling $650 billion, were made with mobile devices (Fung Business Intelligence, 2017). It is projected that by 2019, mobile users in China will spend approximately $1.4 trillion on m-commerce, which will account for 71.5% of the country's total e-commerce sales and nearly a quarter of the country's entire retail market (eMarketer, 2016).

Proliferation of inexpensive smartphones, decreased cost of mobile data, and sophisticated mobile payment apps are fueling the m-commerce boom in China. The Chinese government sees the development of m-commerce as the new engine for the next wave of economic growth and has imposed light policy restraints on retailing and payment on the m-commerce platform (Sun & Chi, 2017).

As one of most popular consumer products, more than 20% of apparel sales in China were made through the online channel in 2015. The online appetite for apparel in China has grown exponentially. Euromonitor International (2016) predicts this trend will continue in the next few years with China contributing half of the world's value growth for online apparel purchases. According to Euromonitor International's 2016 Global Consumer Trends Survey, majority of Chinese consumers preferred the online channel to the in-person experience for both research and purchase of apparel. Owing to the increasingly rapid pace of working and living, digitally savvy Chinese consumers have been seeking more efficient, convenient, and inexpensive ways of shopping, which has resulted in a boost to online purchases.

Apparel m-commerce is a division of the m-commerce channel with a special focus on apparel and related accessories. The trends in apparel online shopping indicate substantial growth in the amount of time spent on retail mobile platforms and overall retail spending on mobile channel (Fulgoni, 2014; Bilgihan, Kandampully, & Zhang, 2016). Apparel retailers are offering greater product acquisition options, competitive prices, and user-friendly shopping experience via their mobile platforms to attract more consumers with the purpose of m-commerce revenue generation. However, the outcomes do not always turn out as expected. According to Fung Business Intelligence (2017), less than 30% of apparel retailers are currently making profits from their m-commerce channel in China. The popularity contrast among apparel mobile retailers (m-tailers) is distinct. Some apparel m-tailers can attract millions of visitors to their web sites everyday while others seem not appealing to consumers.

Park, Jeon, and Sullivan (2015) state that in brick-and-mortar channel, store image and interior quality have an enormous impact on consumer apparel buying decisions, and apparel online retail web sites are similarly important as web site quality affects consumer satisfaction and consequently purchase intention. Despite its importance, extant research has yet to focus on apparel mobile web site quality and its effect on positive consumer shopping experience and future purchase stimulus (Sanakulov & Karjaluoto, 2015). Thus, in order to gain a better understanding of these critical issues regarding apparel m-commerce, this study aims to empirically determine the effects of key features of apparel mobile web site on the Chinese consumer satisfaction that in turn affects their intentions to shop apparel via mobile web sites. Specifically, the objectives of this study are threefold: first, through identifying the key features of apparel mobile web site, this study proposes a research model illustrating the relationships between apparel web site quality, consumer satisfaction, and consumer purchase intention; second, the psychometric properties of the developed model are examined using the primary consumer survey data gathered in China. The significant factors influencing Chinese consumer satisfaction and further intent to use apparel m-commerce are statistically determined. Finally, based on the results, some managerial implications for marketers and companies are provided.

The remainder of this article is organized as follows. The next section reviews the relevant literature and proposes the hypotheses. The research model is then introduced with the corresponding measures and scales for each construct in the model. In the methodology section, the survey subjects, data sets, and statistical methods are described, respectively. The results and discussion follow thereafter. Next, the conclusions are drawn based on the findings, and the implications for both academic researchers and industrial practitioners are presented. Finally, some limitations of this study are addressed, and some directions for future research are offered.

Literature Review and Hypothesis Development

In this section, we review the relevant literature on the theory of self-regulatory process (Bagozzi, 1992) justifying the relationship between m-commerce web site quality, consumer satisfaction, and purchase intention. The measures of mobile web site quality include web site visual appeal, apparel visual appeal, brand trust, web site informativeness, web site response time, and web site security. The hypotheses are proposed based on the review of the literature.

The Theory of Self-Regulatory Process

Theory of self-regulatory process proposed by Bagozzi (1992) explains the relationship between consumer attitude and behavioral intention. Previous attitude theories, such as the theory of reasoned action (Ajzen & Fishbein, 1977) and the theory of planned behavior (Ajzen, 1991), suggested that consumer attitudes could influence their intentions directly. However, Bagozzi (1992) argued that attitudes did not determine intentions sufficiently. He suggested that self-regulatory processes were needed to predict intention.

Bagozzi (1992) defines attitude as an appraisal of the consequences of acting or simply as an appraisal process. He argues that the appraisal process would stimulate emotional reactions, which, in turn, lead to coping response of intention. Therefore, attitude does not influence intention directly. Emotion could mediate the relationship between consumer attitude and behavioral intention. Appraisal would lead to particular emotion, such as satisfaction. In this situation, this emotion will lead to specific intentions, such as to maintaining or increasing satisfaction (Tsai & Bagozzi, 2014). In contrast, if a consumer experiences an unpleasant event or fails to achieve a desired goal, an outcome-desire conflict will occur. Particular emotion, such as dissatisfaction, results. This emotion, in turn, stimulates the intention of avoidance.

Self-regulatory process has been widely applied in consumer behavior research in the context of e-commerce or m-commerce. Barnes and Vidgen (2001), DeLone and McLean (2003), Loiacono, Watson, and Goodhue (2007), and Shin, Chung, Oh and Lee (2013) identified a wide array of web site features that could affect consumers' evaluation of web site quality and result in positive or negative emotion. These investigated web site quality features include informational fit-to-task; tailored communications; trust; response time; ease of understanding; intuitive operations; visual appeal; innovativeness; emotional appeal; consistent image; online completeness. Polites, Williams, Karahanna, and Seligman (2012) used the theory of self-regulatory process as a framework to explore the antecedents of e-satisfaction and the relationship between e-satisfaction and web site loyalty. The result showed that satisfaction mediated the relationship between web site stickiness and consumer's appraisal of information quality, system quality, usefulness, and perceived value of the web site.

Measure of Web site Quality

Web site quality is defined as a multi-dimensional interface, stimulating positive or negative user attitude that results from the interaction between user and web site (Gao & Bai, 2014; Kim & Niehm, 2009). Variables to determine overall m-commerce web site quality were recently introduced to research through the development of the WebQual scale (Loiacono, 2000). Expanding from the Technology Acceptance Model (TAM), WebQual offers an applicable scale that allows researchers to systematically assess the quality of m-commerce web site (Chang & Wang, 2011; Loiacono et al., 2007). TAM was established to measure consumer acceptance of new technology through examination of two primary variables, namely the perceived ease of use and perceived usefulness (Davis, 1989). TAM provides a useful tool to help understand the general consumer perception of new technology and future intent to use the technology. The WebQual scale has expanded the TAM consumer measurements through introducing new constructs to more accurately identify the effects of web site quality on consumer reuse intention. The constructs originally introduced include ease of understanding and intuitive operations related to ease of use, information quality and tailored communications related to gathering information, functional fit-to-task, trust, response time, consistent image, online completeness, relative advantage, and customer service related to carrying out transactions, and visual appeal, innovativeness, and emotional appeal related to entertainment (Loiacono, 2000; Loiacono et al., 2007).

The WebQual constructs measuring web site quality have been applied to the analysis of business-to-consumer (B2C) web sites, which represent the major online interaction between apparel retailers and consumers. Kim and Niehm (2009) indicated among the original WebQual constructs, higher correlations between certain constructs suggested a need for a number of second-order constructs. In a study examining the dimensionality of web site quality for apparel retailers, using the WebQual scale, Kim and Stoel (2004) identified that six constructs of web site quality were highly related to consumer reuse intention. They are web appearance, entertainment, informational fit-to-task, transaction capability, response time, and trust. The second-order constructs often measure the key features of a web site quality (Loiacono et al., 2007) and provide an updated empirical support for m-commerce consumer research (Hasanov & Khalid, 2015).

Entertainment: Web site Visual Appeal and Apparel Visual Appeal

Visual cues are a significant influence among consumers within the apparel industry. Consumers pursuing apparel products perceive desire through the visual displays of clothing and store layout, and make consumption decisions based on these visual sensory details (Khakimdjanova & Park, 2005; Ji & Pang, 2006). Prior studies on apparel online shopping indicate that visual appeal is an important influencing factor and can determine the business success that companies may achieve (Ha & Lennon, 2010). In online shopping, this appeal is described by web site balance, aesthetics, consistency, and response (Gao & Bai, 2014). Visual aspects are powerful in dictating consumer satisfaction and intent to purchase when elicited and experienced by the consumer. In a study by Cyr, Head, Larios, and Pan (2006), mobile web site interface design significantly affected whether a person stays to continue shopping or not. Visual displays on a mobile apparel web site therefore must entice the visitors who integrate visual web site appeal with perceived web site quality.

The apparel industry also relies heavily on imagery to promote the quality of goods produced for sale. While visual merchandising plays a significant role in brick-and-mortar shopping, the effect of visual merchandising in mobile apparel shopping is often overlooked. Moayery, Zamani, and Vazifehdoost (2014) proved that visual merchandising in brick-and-mortar shopping influenced impulsive buying behaviors among female consumers while younger consumers were more responsive to visual merchandising techniques. Visual merchandising in online apparel shopping was revealed to influence consumer pleasure and arousal, among both online shoppers and browsers (Ha & Lennon, 2010).

An empirical study by Khakimdjanova and Park (2005) proved a need for improved visual merchandising of apparel in online commerce applications. Researching the variables of apparel presentation and aesthetics online, the study found that online retailers needed to create a setting similar to brick-and-mortar apparel shopping to achieve the best shopping results. In another study, Flavian et al. (2009) affirmed that web site design is a key factor for apparel retailers to obtain positive outcomes as it significantly affects users' perceptions and behaviors. In order to improve the online shopping experience, apparel visual display and web site structure and design were two important areas that demand attention and development. Therefore, we propose the following hypotheses 1 and 2.

H1: Mobile web site visual appeal positively affects Chinese consumers' web site satisfaction.
H2: Apparel visual appeal positively affects Chinese consumers' web site satisfaction.

Ease of Use: Perceived Information Quality

Online shopping provides new channels for consumers to experience visual and information stimuli, which supports or discourages consumption intentions through product search. One main disadvantage of m-commerce channel is the lack of information quality sought by consumers. Maity and Dass (2014) explored the retailer to consumer communication process across the three shopping channels (i.e., brick-and-mortar stores, e-commerce, and m-commerce). Low media richness was observed among m-commerce web sites, which included limitation in screen size and communication, and influenced the superiority of desktop computer-based e-commerce shopping channel over the m-commerce alternative.

In a study by Kim and Stoel (2004), research asserted that informativeness of an m-commerce web site was a notable factor of predicting consumer satisfaction. Design quality of a mobile web site was said to influence consumer traffic, whereas information quality encouraged purchase (Kim & Stoel, 2004). According to Hausman and Siekpe (2009), acceptance of a web site depends on informativeness, which is viewed as a human factor influencing overall purchase intention. Perceived value of a mobile web site is also a result of provided product information (Kim & Niehm, 2009). Thus, the following hypothesis is proposed.

H3: Information quality on mobile web site positively affects Chinese consumers' web site satisfaction.

Response Time: Response Time of Web site Applications

In order for consumers to perceive a web site useful in making transactions, response time must be factored as a variable for web site quality. Web site response time is defined as the delay a consumer experiences during the attempt to observe online information (Galletta, Henry, McCoy, & Polak, 2006). In the study examining the impact of web site delay on consumer satisfaction, Galletta et al. (2006) asserted that web site response delay and unfamiliarity with a web site had a greater impact on decreasing consumer patience and intent to return. Response time was applied to apparel m-commerce through the study by Kim and Stoel (2004), where response time was proven to significantly affect the consumer perception of transaction-related web site qualities. In this study, satisfaction with an apparel m-commerce web site was related to transaction capabilities. Therefore, we propose the following hypothesis.

H4: Response time positively affects Chinese consumers' web site satisfaction.

Web site Security

Web site security involves protecting consumers from fraud and monetary losses (Bressolles et al., 2014). Loiacono et al. (2007) related consumer transactions to trust regarding the web site processes that influence reuse intention. The concept of trust was developed to reflect the outcome from consumers' perceived safety and privacy of web site processes. Web site security has become one of the most important values desired by consumers during online shopping due to the recent lack of protective measures concerning retail customer security and privacy.

In a study conducted by Chen and Kao (2010), web site security and privacy measures were indicated to be most influential among online consumers in Taiwan. To demonstrate acceptable security and privacy protection, individuals in the study noted that control over the transaction process can improve perceived security and privacy. Complex online interactions were suggested to maximize feelings of security when shopping online, which provided security as a main influence on online shopping satisfaction (Chen & Kao, 2010).

In December 2013, Target, a big box retailer and one of the current leaders in online shopping, was hacked by outside sources (Isidore, 2014). During this event, Target lost details regarding monetary (credit card) and personal information of consumers who shopped during the holiday season. Shortly after the Target credit card breach, security hacks hit other major retailers, including Neiman Marcus, Michaels, and the Home Depot. The security hacks were made public through news releases, instilling fear among consumers regarding online and in-store security measures (Smith, 2014). Following these events and based on the findings from prior studies, we suggest that security features among apparel m-commerce retailers gained importance among consumers. This leads to the hypothesis below.

H5: Security features of mobile web site positively affect Chinese consumers' web site satisfaction.

Brand Trust: Previously Established Consumer Brand Trust

Brand trust is described as the relationship developed between consumers and retailers as a result of learned expectations the consumer relates to the brand (Jones & Kim, 2010). This relationship is often associated with visual cues the consumer interprets when shopping online, and affects the consumer's intention to shop at a particular retailer's web site (Loiacono, 2000). The WebQual scale most closely relates brand trust to consistent image; however, research has proven a need for deeper perspective than given by the WebQual scale. As digital-only stores become more prevalent in online shopping, developing brand trust can be problematic for consumers. Brick-and-mortar stores have a greater chance at developing brand trust in person, transitioning this consumer connection through a well-built online channel. Hahn and Kim (2009) claimed that consumers, who had previously developed trust through a brick-and-mortar retailer experience, could more easily accept the retailer's online shopping channel to meet their needs. While brick-and-mortar stores have greater ability to develop brand trust in their stores, online retailers face problems developing trust.

To influence shopping perceptions, retailers depend on the development of brand trust. According to Shiue and Li (2013), consumers who perceived a brand positively through levels of previous commitment were more likely to show positive attitude and reuse the web site. Ling, Chai and Piew (2010) studied brand orientation and discovered that stronger brand names and images led to greater levels of consumer satisfaction with his/her purchase decisions. Higher levels of comfort associated with brand trust can therefore influence consumer perceptions of web site quality, leading to the hypothesis proposed below.

H6: Previously developed brand trust positively affects Chinese consumers' web site satisfaction.

Consumer Web site Satisfaction and Intent to Use

Web site satisfaction can be the result of different determining factors with changes of web site application (Schaupp, 2010). In a study of impact of web site satisfaction on consumer web site reuse intention, Schaupp (2010) found that web site satisfaction must be met before consumers intended to return.

Based on the theory of self-regulatory process, Cristobal, Flavián and Guinaliu (2007) identified a wide array of web site features including ease of use, response time, visual appeal, and trust that could affect consumers' evaluation of web site quality and result in satisfaction or dissatisfaction outcome. In a similar fashion, the empirical findings from Chen and Kao (2010) supported the web site quality–satisfaction–behavioral intention relationship among Taiwanese consumers. Furthermore, Wakefield, Wakefield, Baker, and Wang (2011) indicated that consumers related satisfaction to an enhanced web site interface and showed greater intent to reuse the web site if satisfaction was achieved. Following the findings of prior relevant studies, we propose the hypothesis below.

H7: Consumer web site satisfaction positively affects future intent to use among Chinese consumers.

Proposed Research Model

Based on the extensive literature review above, a conceptual model including all the proposed relationships is illustrated in Fig. 1. Perceived apparel mobile web site quality affects consumers' satisfaction, which in turn influences their intent to purchase.

Web site quality constructs were derived from the WebQual scale that has been widely applied in prior studies. Specifically, the scale for web site visual appeal was adapted from Wang and Liao (2008). Apparel visual appeal was measured by the

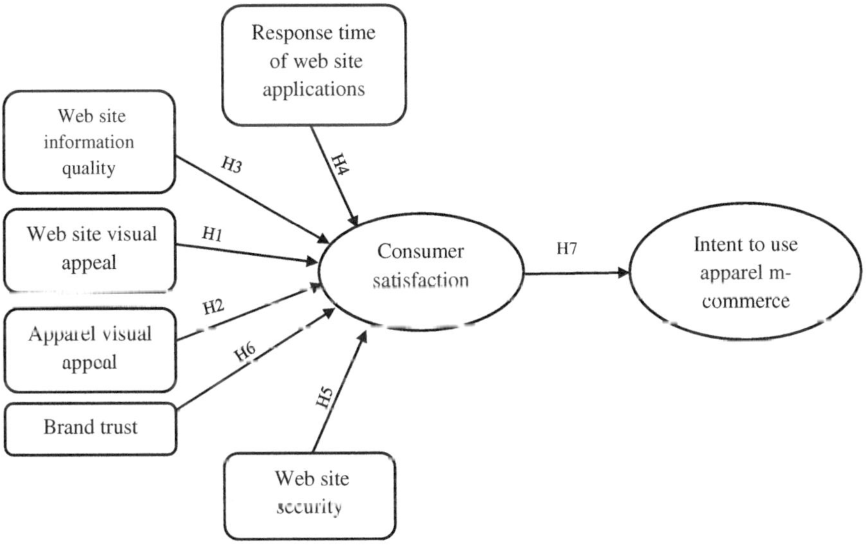

Fig. 1 Proposed mobile web site quality–satisfaction–behavioral intention model

scale developed by Khakimdjanova and Park (2005). The scale for web site response time was adapted from Kim and Stoel (2004). The scale for web site informativeness was adapted from Kim and Niehm (2009). Web site security was captured by the scale from Bressolles, Durrieu, and Giraud (2007). The scale for brand trust was adapted from Jones and Kim (2010). The scale for consumer satisfaction was adapted from Zhou (2011). The scale for intent to use apparel m-commerce was adapted from Hausman and Siekpe (2009). Appendix 1 lists all the constructs and their corresponding measurement scales.

Methodology

Survey Instrument Development and Data Collection

The survey instrument was initially developed in English and reviewed by four professors who were familiar with research topic and survey techniques. Then, the instrument was translated into Chinese by a bilingual graduate student. The Chinese version was back-translated into English by another bilingual graduate student to ensure translation equivalence. The Chinese version of survey instrument was tested through a pilot study with eight Chinese consumers who have made purchases through apparel mobile web sites. The suggestions from the participants were used to refine the instrument with regard to arrangement, wording accuracy, and relevance (Chi, Kilduff, & Gargeya, 2009). This process helped to make the final survey instrument more valid and clearer (Mariadoss, Chi, Tansuhaj, & Pomirleanu, 2016).

The primary data was collected via an online survey of Chinese consumers in January 2017. The professional survey web site used is Sojump (http://www.sojump.com/, similar to Amazon Mechanical Turk https://www.mturk.com/mturk/), which owns a national consumer database in China. A total of 296 eligible responses were received. The profile of survey respondents is presented in Table 1.

Of 296 respondents, 40.3% were female and 59.7% were male. The ages of the respondents varied from 18 years old to over 50 years old, mainly distributing (91.7%) in the range from 18 to 40 years old. Most of the respondents had bachelor's degree (52.3%), followed by some college (29.5%), high school (10.5%), master's degree (6.4%), and doctorate (1.3%). 45.2% of the respondents' personal pre-tax income ranged from ¥50,001 to ¥100,000 (approximately US$7,300–$14,599), followed by ¥20,001 to ¥50,000 at 21.6% (approximately US$2,920–$7,299), ¥100,001 to ¥150,000 at 17.6% (approximately US$14,600–$21,898), ¥150,001 to ¥200,000 at 10.8% (approximately US$21,899–$29,197), less than ¥20,000 (US$2,919) at 3.8%, and more than ¥200,000 (US$29,198) at 2.5%.

Table 1 Profile the survey respondents

	Percentage (%)		Percentage (%)
Gender		*Education level*	
Male	40.3	High school	10.5
Female	59.7	Some college	29.5
		Bachelor's degree	52.3
Age		Master's degree	6.4
18–25	26.5	Doctorate	1.3
26–30	23.8	*Personal annual income level (pre-taxation)*	
31–35	26.1	≤ ¥20,000	3.8
36–40	15.3	¥20,001–¥50,000	20.1
41–45	4.5	¥50,001–¥100,000	45.2
46–50	2.3	¥100,001–¥150,000	17.6
51 or older	1.5	¥150,001–¥200,000	10.8
		>¥200,000	2.5
Annual expenditure on apparel		*Annual expenditure on apparel* via *mobile web sites*	
≤¥1,000	12.5	≤¥1,000	24.6
¥1,001–¥5,000	31.2	¥1,001–¥5,000	49.5
¥5,001–¥10,000	43.1	¥5,001–¥10,000	20.9
¥10,001–¥25,000	6.3	¥10,001–¥25,000	3.0
¥25,001–¥50,000	5.1	¥25,001–¥50,000	1.3
>¥50,000	1.8	>¥50,000	0.7

Note total eligible responses are 296. US$1 = ¥6.85 at the time of data collection

With regard to annual total expenditure on apparel, 43.1% respondents indicated they spent between ¥5,001 and ¥10,000 (approximately US$730–$1,460), followed by 31.2% at a range of ¥1,001–¥5,000 (approximately US$146–$729), 12.5% at ¥1,000 or less (<US$146), 6.3% at a range of ¥10,001–¥25,000 (approximately US $1,461–$3,650), 5.1% at a range of ¥25,001–¥50,000 (approximately US$3,651–$7,299), and 1.8% with ¥50,000 or more (>US$7,299). In contrast, the respondents generally spent less on apparel via mobile web sites. Approximately 95% of the respondents indicated they spent no more than ¥10,000 (>US$1,460) on apparel m-commerce with 49.5% at ¥1,001–¥5,000 (approximately US$146–$729), 24.6% with less than ¥1,000 (<US$146), and 20.9% at ¥5,001–¥10,000 (approximately US$730–$1,460).

Data Analysis Methods

The statistical assumptions including multivariate normality, multicollinearity, and correlations were first examined. Multivariate normality is when each variable under consideration is normally distributed with respect to each other variable. Multivariate normality can be assessed through calculating skewness and kurtosis for each variable. There is a violation of normality assumption, if the results are greater than +1.0 or smaller than −1.0. Multicollinearity is a statistical phenomenon in which two or more predictor variables in a multiple regression model are highly correlated, meaning that one can be linearly predicted from the others with a non-trivial degree of accuracy. In order to test for multicollinearity among the predictor variables, variance-inflation factor (VIF_j) and tolerance were applied. The multicollinearity statistics show that the tolerance indicator should be all greater than 0.2, and their VIF_j values should be less than 5.0 to indicate that no multi-collinearity problem occurs (Ott, Longnecker, & Ott, 2001). Since each construct was measured by multiple items, the average score of the multi-items for a construct was computed and used in further analysis such as correlation analysis and multiple regression analysis (Chi & Sun, 2013). Pearson's correlation analysis was applied to examine the relationship between the constructs. The correlation coefficient value (r), range from 0.10 to 0.29, is considered weak, from 0.30 to 0.49 is considered medium, and from 0.50 to 1.0 is considered strong (Chi & Zheng, 2016). According to Field (2009), correlation coefficient should not go beyond 0.8 to avoid multi-collinearity. Exploratory factor analysis (EFA) and confirmatory factor analysis (CFA) were employed to test the constructs in the proposed model in terms of reliability, unidimensionality, and construct validity including both convergent validity and discriminant validity.

For factor analysis, the extraction criterion is set as eigenvalue above 1.0. Items with low factor loadings (less than 0.50) are dropped (Yoo & Donthu, 2001). The deduction of certain measurement variables requires the re-examination of factor loadings, coefficient alpha, item-to-total correlations, and factor structure (Chi et al., 2009). This iterative procedure is repeated until all requirements are met.

Unidimensionality, reliability, convergent validity, and discriminant validity were tested for proving model adequacy. First, unidimensionality refers to the existence of one underlying measurement construct that accounts for variation in examinee responses. Second, Cronbach's alpha is a coefficient of internal consistency. It is commonly used as an estimate of the reliability of a psychometric test for a sample of examinees. Third, convergent validity refers to the extent to which indicators of a specific construct "converge" or share a high proportion of variance in common. Convergent validity is valid when average variance extracted (AVE) scores for all latent constructs are above the desired threshold of 0.50. AVE is a summary measure of convergence among a set of items representing a construct. It is the average percent of variation explained among the items. Fourth, discriminant validity refers to the extent to which a construct is truly distinct from other constructs. Comparing the AVEs to the squared correlation between the two latent constructs of interest, the AVEs should be greater than the squared correlation in order to demonstrate satisfactory discriminant validity (Nunnally & Bernstein, 1978). If the AVEs by the correlated latent constructs are greater than the square of the correlation between the latent constructs, discriminant validity is obtained (Fornell & Larcker, 1981).

Multiple regression method is applied to analyze the relationship between dependent variable and independent variable(s) and test the hypotheses. SPSS software was used for statistical assumption tests, model adequacy examinations, and multiple regression analysis.

Results and Discussions

Statistical Assumption Examination, Factor Analysis, and Construct Adequacy

Table 2 presents correlations and properties of all constructs. All skewness and kurtosis scores are between +1.0 and −1.0, which suggest there are no violations of normality assumption. All VIF values are below five, suggesting there are no

Table 2 Correlations and properties of all constructs

	WIQ	RT	S	T	WVA	AVA	SA	IU
WIQ	1	0.102	0.132*	0.236**	0.262**	0.165**	0.337**	0.254**
RT	*0.010*	1	0.375**	0.336**	0.284**	0.246**	0.358**	0.273**
S	*0.017*	*0.141*	1	0.412**	0.218**	0.270**	0.468**	0.509**
T	*0.056*	*0.113*	*0.170*	1	0.253**	0.187*	0.531**	0.524**
WVA	*0.069*	*0.081*	*0.048*	*0.064*	1	0.389**	0.318**	0.305**
AVA	*0.027*	*0.061*	*0.073*	*0.035*	*0.151*	1	0.244**	0.188**
SA	*0.114*	*0.128*	*0.219*	*0.282*	*0.101*	*0.060*	1	0.567**
IU	*0.065*	*0.075*	*0.259*	*0.275*	*0.093*	*0.035*	*0.321*	1
Mean	4	4	4	4	4	4	4	4
S.D.	0.5	0.4	0.5	0.4	0.6	0.5	0.4	0.4
Cronbach's alpha	0.880	0.847	0.913	0.877	0.735	0.710	0.912	0.948
AVE	0.501	0.714	0.573	0.520	0.500	0.503	0.505	0.680
Chi-square test *p* value	0.085	0.105	0.093	0.108	0.121	0.096	0.075	0.132
Skewness	0.435	0.126	0.233	0.256	−0.165	−0.271	0.250	0.146
Kurtosis	0.312	0.167	0.211	0.336	0.223	0.242	0.462	0.224
VIF	1.853	1.781	1.725	1.664	1.815	1.462	1.887	1.706

Note the italic numbers are the squared corresponding correlations. *: correlation is significant at the 0.05 level (2-tailed). **: correlation is significant at the 0.01 level (2-tailed)

multicollinearity issues among constructs and variables. After exploratory factor analysis, the measurement variables labeled as WVA3, AVA2, AVA5, T5, T8, T10, and T11 were dropped due to low factor loading (see Appendix 1). All the factor loadings of the remaining measurement items to their respective constructs are high (0.6 and higher) and statistically significant, while their loadings to other constructs are very low (0.3 and lower). This also shows unidimensionality for the constructs. In addition, the chi-square tests of all constructs were insignificant, which established the evidences of unidimensionality. Cronbach's alphas of all constructs except compatibility are greater than 0.70, indicating reliability is rigorously met (Nunnally & Bernstein, 1978). The AVE scores for all constructs are above the desired threshold of 0.50, suggesting convergent validity. All AVE scores are greater than the squared corresponding correlations, which demonstrate satisfactory discriminant validity.

Hypothesis Testing Results and Discussion

Once the adequacies of all constructs were demonstrated, the proposed hypotheses were tested using multiple regression technique. A single score was obtained for each construct by averaging across the measurement items. Table 3 presents the results of hypothesis testing. Among seven hypotheses, five of them (H1, 2, 3, 5, and 7) were statistically significant at a $p < 0.05$ level and two (H4 and H6) were insignificant. The effects of demographic variables on consumer satisfaction toward apparel mobile web site and their intent to use apparel m-commerce were all insignificant at a $p < 0.05$ level.

Figure 2 illustrates the identified relationships in the proposed research model. Web site information quality, web site visual appeal, apparel visual appeal, and web site security positively affect Chinese consumer satisfaction toward apparel mobile web sites, while the impacts of response time of web site applications and brand trust on consumer satisfaction are insignificant. There are no significant differences between age groups, genders, education levels, or income levels among Chinese consumers in regard to their satisfaction toward apparel mobile web sites. These constructs and demographic factors collectively exhibit a good explanatory power, accounting for 52.6% of variance in consumer satisfaction of apparel mobile web sites. Consequently, consumer satisfaction of apparel mobile web sites positively affects their intent to purchase apparel through the mobile web sites and accounts for 37.5% of variance in purchase intention. Consumer satisfaction shows a good predictability for their behavioral intention.

Table 3 Results of hypothesis testing

Hyp.		DV	IDV	Std. Coef. (β)	t-value	Sig. at $p < 0.05$	Control variable	Std. Coef. (β)	t-value	Sig. at $p < 0.05$	Total R^2	F-value (df1/df2)	Sig. at $p < 0.05$
		SA	Constant	–	2.618	0.025	Age	−0.037	−0.824	0.411	0.526	30.62 (10/285)	<0.001
H1	Y		WVA	0.267	3.365	0.007	Gender	0.008	0.076	0.828			
H2	Y		AVA	0.123	2.074	0.036	Education	−0.056	−1.124	0.262			
H3	Y		WIQ	0.210	2.715	0.006	Income	−0.042	−0.558	0.470			
H4	N		RT	0.011	0.152	0.776							
H5	Y		S	0.338	5.403	<0.001							
H6	N		T	0.096	1.634	0.083							
		IU	Constant	–	5.435	<0.001	Age	0.032	0.654	0.513	0.375	26.55 (5/290)	<0.001
H7	Y		SA	0.602	9.884	<0.001	Gender	0.095	1.796	0.062			
							Education	0.059	1.218	0.253			
							Income	0.027	0.784	0.485			

Note: *Y* hypothesis supported; *N* hypothesis not supported; *IU* intent to use; *WVA* web site visual appeal; *AVA* apparel visual appeal; *WIQ* web site information quality; *RT* response time of web site applications; *S* web site security; *T* brand trust; *SA* consumer satisfaction. *Std. Coef.* stands for standardized coefficients; *DV* dependent variable; *IDV* independent variable

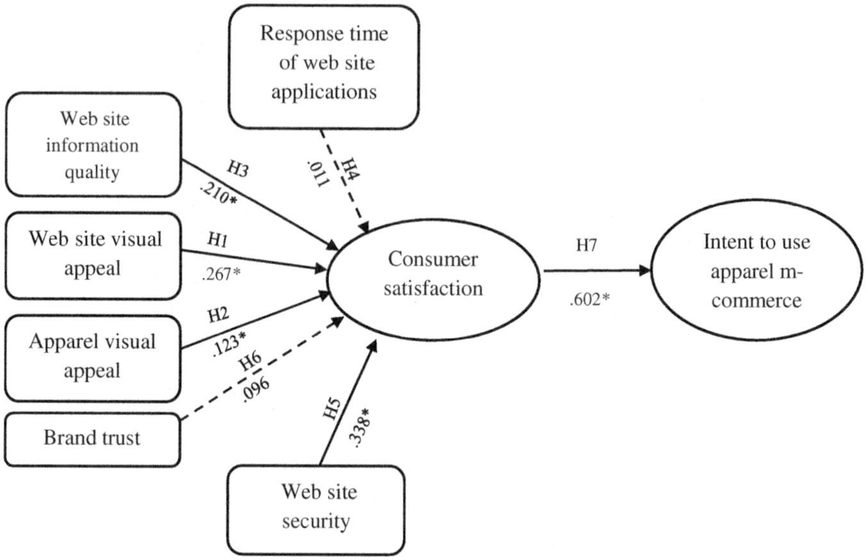

Fig. 2 Identified relationships in the research model. *Note* * statistical relationship is significant at $p < 0.05$ level

Conclusions

As one of the first efforts made to understand the emerging apparel m-commerce phenomenon in China, this study empirically determined the effects of the key features of apparel mobile web site on consumer satisfaction and consequently the impact of consumer satisfaction on their intent to use apparel mobile commerce. The findings reveal how an apparel mobile web site should be designed to enhance Chinese consumer satisfaction that in turn leads to purchase outcome. The applicability and usefulness of the WebQual scale were proven with the adaptation to the apparel context.

Overall, the main contributions of this study to the existing body of literature are threefold. First, building on the theory of self-regulatory process, this study proposed a mobile web site quality–satisfaction–behavioral intention model. The model shows a high explanatory power for the Chinese consumer satisfaction toward apparel m-commerce web site (R^2 = 52.6%) and demonstrates the significant relationship between consumer satisfaction toward mobile web site and their intent to use apparel m-commerce. Second, the psychometric properties of the proposed model were examined using the gathered primary consumer survey data. All the latent constructs met the criteria of unidimensionality, reliability, and construct validity. The significant factors influencing the Chinese consumer satisfaction toward apparel m-commerce web site include web site information quality, web site visual appeal, apparel visual appeal, and web site security.

Finally, although the investigated demographic factors did not impose statistically significant impacts on Chinese consumer satisfaction toward apparel m-commerce web site, some noteworthy patterns between consumer segments emerged. Chinese consumers with younger age, male, lower education level, or lower income level tend to show a higher satisfaction toward apparel m-commerce web site. Meanwhile, among Chinese consumers who show a higher satisfaction toward apparel m-commerce web site, those who are older, male, higher education level or higher income level are more likely to make actual purchases via apparel m-commerce web site.

Implications

This study also imparted some managerial implications. Given the accelerating competition facing apparel m-tailers in China, attracting more consumers to visit their web sites and enhancing their satisfaction toward the web sites that leads to use and reuse have become crucial. Based on the consumer desired features of an apparel m-commerce web site identified in this study, apparel m-tailers can create or modify their web sites with clear direction and targeted outcomes. Investment decision may be made in a more effective manner.

Mobile web site interface design with balance, aesthetics, and consistency appeals to Chinese consumers. Given the characteristics of apparel, visual merchandising plays a significant role in apparel online shopping, which influences consumer pleasure and arousal. M-tailers need to create a setting similar to brick-and-mortar apparel shopping to achieve the best shopping results. Apparel presentation and aesthetics on the mobile web site should promote the quality of goods. Compared to desktop computer-based e-commerce channel, low media richness has been one of prominent disadvantages for m-commerce web sites due to the limitation in screen size, although recent popularity of big screen smartphones partially addresses the issue. Providing relatively simple but sufficient product and service information for consumers to make purchase decision is the principle for any m-commerce web design. Consumer trust is the outcome from consumers' perceived safety and privacy of web site processes. Web site security has become one of the most important values desired by consumers during online shopping due to the recent lack of protective measures concerning retail customer security and privacy. As lack of web site security is fatal for any m-tailers, investment in web site security should always be one of top priorities.

With the rapid proliferations of 4G network and decreased cost of mobile data in China, web site response time is no longer a concern to Chinese consumers. The continuous upgrading of mobile Internet infrastructure is one of the major factors for the recent boom of m-commerce in China. According to a recent study from Fung Business Intelligence (2017), most of m-commerce users in China are satisfied with the speed of mobile Internet connection and business transaction. Compared to the popularity of selling through their own mobile web sites by apparel brands and retailers in other major countries such as USA, a majority of apparel companies in China, particularly small- and medium-sized enterprises (SMEs), sell through some national reputable digital online platforms such as Tmall.com and JD.com. A great percentage of Chinese m-commerce users are used to search and find what they want through these popular platforms. These platforms usually provide additional product quality warranty to shoppers through their agreements with the product suppliers. This may explain that brand trust does not significantly affect Chinese consumer satisfaction toward apparel m-commerce web sites, which lowers the m-commerce entry barrier for SMEs.

Moreover, based on the findings of this study, certain types of consumers are worth more attention from companies or should be weighed more in company's m-commerce marketing strategies or investment. First of all, younger, male, lower education level, or lower income Chinese consumers tend to be more receptive to the new shopping channel—m-commerce that is innovative, private/personal, convenient, price-competitive and requires less resources compared to other shopping channels. Apparel m-tailers can rapidly expand customer base through attracting these consumers. Second, for apparel m-tailers to develop long-term customer base, older, male, higher education level or higher income consumers seem to be more loyal and are more likely to make actual purchases once they are satisfied with their experiences with apparel m-commerce web sites. Finally, with the rapid development of m-commerce in China, the market is becoming saturated. Targeting those consumers who are more challenging to be pleased and have higher expectation on shopping experience could help apparel m-tailers gain market share and create unique market niche.

Limitations and Future Studies

Although this study has provided a better understanding on the effects of the key features of apparel m-commerce web site on consumer satisfaction and the impact of consumer satisfaction on their intent to use apparel m-commerce in China, there are some limitations that should be mentioned and could be addressed in the future studies. First, this study used a quantitative research method. Although the quantitative approach allows the examination of the casual relationships between independent factors and consumer satisfaction and consequently intent to use apparel m-commerce, it is considered weak in discovering the underlying reasons to the phenomenon. Qualitative methods used in future studies might provide a more detailed reasoning about the relationships identified in the quantitative analysis. Second, additional factors could be added to the model in the future studies to achieve a higher explanatory power on consumer intent to use apparel m-commerce. Finally, although the technique is valid and has been applied extensively in the prior studies, the use of a single indicant for multi-item latent construct in data analysis could lose certain information about the construct. In the future, other methods such as structural equation modeling could be applied for data analysis.

Appendix 1: Constructs and corresponding measurement scales

Construct	Measure and scale [Factor loading]	Source
Intent to use (IU)	IU1: I will definitely buy apparel from this mobile web site in the near future. [0.821] IU2: I intend to purchase apparel through this mobile web site in the near future. [0.815] IU3: It is likely that I will purchase apparel through this mobile web site in the near future. [0.786] IU4: I expect to purchase apparel through this mobile web site in the near future. [0.873]	Hausman and Siekpe (2009)
Consumer satisfaction usefulness (SA)	SA1: I feel satisfied with the services provided by this mobile web site. [0.691] SA2: I feel contented with the services provided by this mobile web site. [0.687] SA3: I like the services provided by this mobile web site. [0.753]	Zhou (2011)
Web site information quality (WIQ)	WIQ1: The mobile web site provides accurate information. [0.769] WIQ2: The mobile web site is informative. [0.751] WIQ3: The mobile web site provides updated information. [0.709] WIQ4: The mobile web site provides high quality information. [0.702] WIQ5: The mobile web site provides timely information. [0.736] WIQ6: The mobile information on the web site is relevant to me. [0.675] WIQ7: I can find what I need in the mobile web site. [0.601] WIQ8: The mobile web site provides relevant information. [0.704]	Kim and Niehm (2009)
Web site visual Appeal (WVA)	WVA1: The mobile web site uses colors properly. [0.735] WVA2: The mobile web site uses fonts properly. [0.754] WVA3: The mobile web site uses multimedia features properly [Dropped due to low factor loading] WVA4: The layout of the mobile web site is appropriate. [0.628] WVA5: The mobile web site looks organized. [0.702] WVA6: The mobile web site is user-friendly. [0.711]	Wang and Liao (2008)
Apparel visual appeal (AVA)	AVA1: Apparel displayed on a mannequin is visually pleasing. [0.783] AVA2: Apparel displayed on a human model is visually pleasing. [Dropped due to low factor loading] AVA3: Viewing apparel color options on mannequin/model is visually pleasing. [0.806] AVA4. White apparel backdrops are visually pleasing. [0.607]	Khakimdjanova and Park (2005)

(continued)

(continued)

Construct	Measure and scale [Factor loading]	Source
	AVA5: Apparel backdrops are visually pleasing. [Dropped due to low factor loading] AVA6: Apparel item coordination is important. [0.670] AVA7: The visual display of apparel items is easy to understand. [0.588] AVA8: Desirable visual display of apparel items includes: 2D product, 360° garment view, picture enlargement, visual color swatch, changeable garment color on mannequin/model. [0.659]	
Brand trust (T)	T1: I trust the retailer owning the apparel mobile web site. [0.660] T2: The retailer's apparel is not consistent in its quality. *[0.692] T3: The retailer's apparel performs consistently. [0.688] T4: I feel secure when I buy apparel from this retailer's mobile web site because I know it will never let me down. [0.556] T5: The retailer is reputed to perform well. [Dropped due to low factor loading] T6: I have heard negative comments about this retailer. * [0.915] T7: This retailer has a reputation for being unreliable. * [0.820] T8: This retailer has a reputation for being good. [Dropped due to low factor loading] T9: I feel I can trust this retailer completely. [0.722] T10: This retailer's apparel performance tends to be quite variable. *[Dropped due to low factor loading] T11: I can't always be sure how it will perform the next time I buy it. *[Dropped due to low factor loading] T12: I can always anticipate correctly how this retailer's apparel will perform. [0.725] T13: When I buy apparel from this retailer's mobile web site I know exactly what to expect. [0.652]	Jones and Kim (2010)
Web response time (RT)	RT1: The mobile web site takes long to load. *[0.842] RT2: The mobile web site loads quickly. [0.880] RT3: When I use the mobile web site there is very little waiting time between my actions and the web site's response. [0.811]	Kim and Stoel (2010)
Web site security (S)	S1: Generally, I have confidence in the web site's security. [0.785] S2: The mobile web site guarantees security in navigation. [0.628] S3: I think my personal details are safe on this mobile web site. [0.837] S4: I have confidence that the mobile web site will not use my personal information for undesirable purposes. [0.762]	Bressolles et al. (2007)

Note Five-point Likert scale from "Strongly disagree = 1" to "Strongly agree = 5". *: Reversed measures. WVA3, AVA2, AVA5, T5, T8, T10, and T11 are dropped due to low factor loading

References

Ajzen, I. (1991). The theory of planned behavior. *Organizational Behavior and Human Decision Processes, 50*(2), 179–211.

Ajzen, I., & Fishbein, M. (1977). Attitude-behavior relations: A theoretical analysis and review of empirical research. *Psychological Bulletin, 84*(5), 888.

Bagozzi, R. P. (1992). The self-regulation of attitudes, intentions, and behavior. *Social Psychology Quarterly, 55*(2), 178–204.

Barnes, S., & Vidgen, R. (2001). An evaluation of cyber-bookshops: The WebQual method. *International Journal of Electronic Commerce, 6*(1), 11–30.

Bilgihan, A., Kandampully, J., & Zhang, T. (2016). Towards a unified customer experience in online shopping environments: Antecedents and outcomes. *International Journal of Quality and Service Sciences, 8*(1), 102–119.

Bressolles, G., Durrieu, F., & Giraud, M. (2007). The impact of electronic service quality's dimensions on customer satisfaction and buying impulse. *Journal of Customer Behaviour, 6*(1), 37–56.

Bressolles, G., Durrieu, F., & Senecal, S. (2014). A consumer typology based on e-service quality and e-satisfaction. *Journal of Retailing and Consumer Services, 21*(6), 889–896.

Chang, H., & Wang, H. W. (2011). The moderating effect of customer perceived value on online shopping behaviour. *Online Information Review, 35*(3), 333–359.

Chen, C. F., & Kao, Y. L. (2010). Relationships between process quality, outcome quality, satisfaction, and behavioural intentions for online travel agencies–evidence from Taiwan. *The Service Industries Journal, 30*(12), 2081–2092.

Chi, T., Kilduff, P. P., & Gargeya, V. B. (2009). Alignment between business environment characteristics, competitive priorities, supply chain structures, and firm business performance. *International Journal of Productivity and Performance Management, 58*(7), 645–669.

Chi, T., & Sun, Y. (2013). Development of firm export market oriented behavior: Evidence from an emerging economy. *International Business Review, 22*(1), 339–350.

Chi, T., & Zheng, Y. (2016). Understanding environmentally friendly apparel consumption: An empirical study of Chinese consumers. *International Journal of Sustainable Society, 8*(3), 206–227.

Cristobal, E., Flavián, C., & Guinaliu, M. (2007). Perceived e-service quality (PeSQ) measurement validation and effects on consumer satisfaction and web site loyalty. *Managing Service Quality: An International Journal, 17*(3), 317–340.

Cyr, D., Head, M., Larios, H., & Pan, B. (2006). Design aesthetics leading to m-loyalty in mobile commerce. *Information & Management, 43*(8), 950–963.

Davis, F. D. (1989). Perceived usefulness, perceived ease of use and user acceptance of information technology. *MIS Quarterly, 13*(3), 319–340.

DeLone, W. H., & McLean, E. (2003). The DeLone and McLean model of information systems success: A ten-year update. *Journal of Management Information Systems, 19*(4), 9–30.

eMarketer (2016). *Ecommerce turns into Mcommerce in China*. Retrieved from https://www.emarketer.com/Article/Ecommerce-Turns-Mcommerce-China/1013736 on April 10, 2017.

Euromonitor. (2016). *Apparel category to drive product-based digital commerce growth: Think with google partnership*. Retrieved from http://blog.euromonitor.com/2016/09/apparel-category-drive-product-based-digital-commerce-growth.html on May 1, 2017.

Field, A. (2009). *Discovering statistics using SPSS*. Sage publications.

Flavian, C., Guinea, R., & Orús, C. (2009). Web design: A key factor for the website success. *Journal of Systems and Information Technology, 11*(2), 168–184.

Fornell, C., & Larcker, D. F. (1981). Structural equation models with unobservable variables and measurement error: Algebra and statistics. *Journal of Marketing Research, 18*(3), 382–388.

Fulgoni, G. M. (2014). Omni-channel retail insights and the consumer's path-to-purchase. *Journal of Advertising Research, 54*(4), 377–380.

Fung Business Intelligence. (2017). *China retail & e-commerce quarterly Issue 1*. Retrieved from https://www.fbicgroup.com/sites/default/files/CREQ_01.pdf on May 15, 2017.

Fung Business Intelligence. (2017). *China retail & e-commerce quarterly Issue 04*. Retrieved from https://www.fbicgroup.com/sites/default/files/CREQ_04.pdf. on December 10th, 2017.

Galletta, D., Henry, R., McCoy, S., & Polak, P. (2006). When the wait isn't so bad: The interacting effects of website delay, familiarity, and breadth. *Information Systems Research, 17*(1), 20–37.

Gao, L., & Bai, X. (2014). Online consumer behaviour and its relationship to website atmospheric induced flow: Insights into online travel agencies in China. *Journal of Retailing and Consumer Services, 21*(4), 653–665.

Ha, Y., & Lennon, S.J. (2010). Effects of site design on consumer emotions: role of product involvement. *Journal of Research in Interactive Marketing, 4*(2), 80-96.

Hahn, K., & Kim, J. (2009). The effect of offline brand trust and perceived internet confidence on online shopping intention in the integrated multichannel context. *International Journal of Retail & Distribution Management, 37*(2), 126–141.

Kim, J., & Stoel, L. (2010). Factors contributing to rural consumers' in shopping behavior: effects of institutional environment and social capital. *Marketing Intelligence & Planning, 28*(1), 70–87.

Hasanov, J., & Khalid, H. (2015). The impact of website quality on online purchase intention of organic food in Malaysia: A WebQual model approach. *Procedia Computer Science, 72,* 382–389.

Hausman, A. V., & Siekpe, J. S. (2009). The effect of web interface features on consumer online purchase intentions. *Journal of Business Research, 62*(1), 5–13.

Isidore, C. (2014). *Target: Hacking hit up to 110 million customers*. Retrieved from http://money.cnn.com/2014/01/10/news/companies/target-hacking/index.html on May 25, 2017.

Ji, X., & Pang, C. (2006). Online visual merchandising of apparel e-retailers. In *2006 IEEE International Conference on Engineering Management* (pp. 307–311).

Jones, C., & Kim, S. (2010). Influences of retail brand trust, off-line patronage, clothing involvement and website quality on online apparel shopping intention. *International Journal of Consumer Studies, 34*(6), 627–637.

Khakimdjanova, L., & Park, J. (2005). Online visual merchandising practice of apparel e-merchants. *Journal of Retailing and Consumer Services, 12*(5), 307–318.

Kim, S., & Stoel, L. (2004). Apparel retailers: Website quality dimensions and satisfaction. *Journal of Retailing and Consumer Services, 11*(2), 109–117.

Kim, H., & Niehm, L.S. (2009). The impact of website quality on information quality, value, and loyalty intentions in apparel retailing. *Journal of interactive marketing, 23*(3), 221–233.

Ling, K., Chai, L., & Piew, T. (2010). The effects of shopping orientations, online trust and prior online purchase experience toward customers' online purchase intention. *International Business Research, 3*(3), 63–76.

Loiacono, E. T. (2000). *Webqual (tm): A web site quality instrument*. University of Georgia.

Loiacono, E., Watson, R., & Goodhue, D. (2007). WebQual: An instrument for consumer evaluation of web sites. *International Journal of Electronic Commerce, 11*(3), 51–87.

Mariadoss, B. J., Chi, T., Tansuhaj, P., & Pomirleanu, N. (2016). Influences of firm orientations on sustainable supply chain management. *Journal of Business Research, 69*(9), 3406–3414.

Maity, M., & Dass, M. (2014). Consumer decision-making across modern and traditional channels: e-commerce, m-commerce, in-store. *Decision Support Systems, 61*(1), 34–46.

Moayery, M., Zamani, S., & Vazifehdoost, H. (2014). Effect of visual merchandising on apparel impulse buying behaviors among Iranian young adult females. *Indian Journal of Science and Technology, 7*(3), 360–366.

Nunnally, J., & Bernstein, I. (1978). *Psychometric theory*. New York: McGraw-Hill.

Ott, L., Longnecker, M., & Ott, R. L. (2001). *An introduction to statistical methods and data analysis*. CA: Duxbury Pacific Grove.

Park, H. H., Jeon, J. O., & Sullivan, P. (2015). How does visual merchandising in fashion retail stores affect consumers' brand attitude and purchase intention? *The International Review of Retail, Distribution and Consumer Research, 25*(1), 87–104.

Polites, G. L., Williams, C. K., Karahanna, E., & Seligman, L. (2012). A theoretical framework for consumer e-satisfaction and site stickiness: An evaluation in the context of online hotel reservations. *Journal of Organizational Computing and Electronic Commerce, 22*(1), 1–37.

Sanakulov, N., & Karjaluoto, H. (2015). Consumer adoption of mobile technologies: A literature review. *International Journal of Mobile Communications, 13*(3), 244–275.

Schaupp, L. (2010). Web site success: Antecedents of web site satisfaction and re-use. *Journal of Internet Commerce, 9*(1), 42–64.

Shin, J. I., Chung, K. H., Oh, J. S., & Lee, C. W. (2013). The effect of site quality on repurchase intention in Internet shopping through mediating variables: The case of university students in South Korea. *International Journal of Information Management, 33*(3), 453–463.

Shiue, Y. C., & Li, S. H. (2013). Brand involvement in retaining customers despite dissatisfaction. *Social Behavior and Personality: An International Journal, 41*(4), 643–650.

Smitch, C. (2014). *Confirmed: Home Depot credit card hack significantly bigger than Target's*. Retrieved from http://bgr.com/2014/09/19/home-depot-vs-target/ on May 15, 2017.

Tsai, H. T., & Bagozzi, R. P. (2014). Contribution behavior in virtual communities: Cogntiive, emotional, and social influences. *MIS Quarterly, 38*(1), 143–163.

Wakefield, R. L., Wakefield, K. L., Baker, J., & Wang, L. C. (2011). How website socialness leads to website use. *European Journal of Information Systems, 20*(1), 118–132.

Wang, Y. S., & Liao, Y. W. (2008). Assessing eGovernment systems success: A validation of the DeLone and McLean model of information systems success. *Government Information Quarterly, 25*(4), 717–733.

Yoo, B., & Donthu, N. (2001). Developing and validating a multidimensional consumer-based brand equity scale. *Journal of Business Research, 52*(1), 1–14.

Zhou, T. (2011). Examining the critical success factors of mobile website adoption. *Online Information Review, 35*(4), 636–652.

Attitudes Toward Technology, Digital Activities, and Use of Internet Shopping Features Among Chinese and US College Students

Jane E. Workman and Seung-Hee Lee

Abstract The purpose of this study was to examine how attitudes toward technology (ATT), habitual use of digital activities, and frequency of use of internet shopping features differ among male and female Chinese and US college students. Participants were 114 male, 92 female students from a Midwestern US university and 100 male, 93 female students from Lanzhou University in China. Results revealed that US and Chinese students shared some attitudes and behaviors (positive ATT, use of internet shopping features) but differed in others (negative and anxious/dependent ATT, habitual use of digital activities). US (vs. Chinese) students indicated more habitual use of all digital activities. Men and women did not differ in positive or negative ATT but women indicated greater anxiety/dependence on technology than men did. Women indicated more habitual use of total digital activities, social media, and smartphone than men. Implications for Chinese consumers and the fashion market are provided.

Keywords Attitudes toward technology · Digital activities · Internet shopping features

Given the investment of companies in e-commerce, understanding the conditions that must exist for consumers to make use of available technology is an important research topic with economic ramifications for both consumers and online stores. There are many interrelated factors that can affect whether a consumer uses, or does not use, available technology to purchase products online: (a) culture, (b) gender, (c) information processing, (d) motivations, (e) attitudes toward technology, and (f) habit. For example, culture and gender affect the way individuals have learned to process information. Cultural values regarding egalitarianism may lead to gender

J. E. Workman (✉) · S.-H. Lee
Fashion Design and Merchandising, Southern Illinois University, 311 Quigley Hall, Carbondale, IL 62901, USA
e-mail: jworkman@siu.edu

S.-H. Lee
e-mail: shlee@siu.edu

Y. Xu et al. (eds.), *Chinese Consumers and the Fashion Market*, Springer Series in Fashion Business, https://doi.org/10.1007/978-981-10-8429-4_4

differences in use of technology. Culture and gender may interact to affect technology use. Gender affects acceptance and use of technology; therefore, it is important to understand the gender differences in hedonic and utilitarian motivations underlying use of internet shopping features. Established attitudes toward technology and habit, or behavior exhibited in the past, are other variables that can help to explain and predict technology use. China and the USA differ considerably in five of six cultural values: individualism, indulgence, long-term orientation, power distance, and uncertainty avoidance but not in masculinity. Therefore, because of these cultural differences, attitudes toward technology, habitual use of digital activities, and use of internet shopping features may differ among a sample of male and female Chinese and US college student consumers.

The unified theory of acceptance and use of technology (UTAUT) confirms a critical need to examine acceptance and use of consumer technologies—a "multi-billion dollar industry given the number of technology devices, applications, and services targeted at consumers" (Venkatesh, Thong, & Xu, 2012, p. 158). Consumer services include internet shopping features. If, or how often, consumers use technology services may depend on their culture, gender, attitudes toward technology, habitual use of digital activities, and characteristics of the internet shopping features. The purpose of this study was to examine how attitudes toward technology, habitual use of digital activities, and frequency of use of internet shopping features differ among a sample of male and female Chinese and US college student consumers.

China has shown rapid technology adoption trends. The statistics in this paragraph were gleaned from the Statistical Report (July 2016) provided by China Internet Network Information Center. In June of 2016, China had 710 million internet users—an increase of 21.32 million from December 2015. Among internet users, the number of mobile internet users in China reached 92.5% (656 million) making the mobile phone the most important device for internet access. Among Chinese internet users, the ratio of males to females is 53% to 47%. Among Chinese internet users, 20.1% are in the age group of 10–19; 30.4% are in the age group of 20–29; and 24.2% are in the age group of 30–39 for a total of 74.7% in the age group of 10–39. Chinese internet users stayed online for an average of 26.5 h a week. Part of that time was spent shopping; in 2016, there were nearly 448 million online shoppers in China, 63% of those with internet access engaged in online shopping—up to 8.3% from 2015. Mobile devices were used to engage in online shopping by 401 million consumers—61% of those with internet access and an increase of 18% from 2015. With these many online shoppers, China's online retail business has shown continuous growth. Yet China, with an internet penetration rate (i.e., percentage of population using the internet) of 51.7% in 2016, has yet to reach its potential online population.

"The rise of China to become the largest e-commerce importer can be attributed in part to growing health concerns, a rise in disposable income, as well as a growing appetite for overseas luxury and fashion items" (Lee, Gillai, Chen, & Rammohan, 2016, p. 4). China was the world's largest online retail market in 2015 with an increase of 33.3% in online sales compared to 2014 (Tong, 2016) and projected to reach a transaction value of $1.56 trillion by 2018 (E-commerce Drives Retail Sales, 2015). China has been transitioning toward a consumption-driven economy and has experienced several cultural trends, technological innovations, and improved logistics networks that have encouraged the growth of e-commerce. For example, China's large and growing middle class becomes more aware of brand names and more loyal to particular brands plus an increasing tendency to purchase high-quality and/or discretionary products (Lee et al., 2016). Many products they look for are easier to find and cheaper online. Further, the prevalence of mobile devices and product reviews posted on social media sites are strong forces influencing e-commerce (E-commerce in China, 2014).

In comparison, the US has an internet penetration rate of 88.5% (Internet Users by Country, 2016). Internet use in the USA is nearly equal among men and women. With nearly 290 million internet users as of 2016, the USA is among the largest online markets worldwide. Online sales in the USA in 2015 were nearly $342 billion (US Census Bureau News, 2016). About 92% of internet users access an online retail store each month (Bose, 2016). Plus, the Internet influenced in-store sales via in-store Web browsing, reading reviews online before making a purchase, checking what items a store has in stock, checking store hours, reserving an item for in-store pickup, scanning items for a quick price check, and allowing the store to match a lower price elsewhere (Martin, 2016; Sterling, 2017). Indeed, 55% of shoppers acknowledged that customer reviews affect their buying decisions; nearly 40% of in-store customers read online reviews before buying a product or service (Internet Trends, 2017). Offline spending influenced by the Internet has been estimated between $1 and $2+ trillion. According to a Forrester research report, $1.26 trillion of local retail sales were affected in some way by digital media in 2016 (Sterling, 2017).

Theoretical Framework

UTAUT evolved from a synthesis of eight technology use theories (Venkatesh et al., 2012). There are four key constructs in UTAUT that are predicted to influence technology adoption and use: performance expectancy (i.e., degree to which using a technology will be beneficial—utilitarian value or extrinsic motivation), effort expectancy (i.e., ease of use), social influence (i.e., importance of the opinions of important others), and facilitating conditions (i.e., perceptions of resources and support available). Further, individual difference variables (e.g., gender, age, experience, and voluntariness of use) are theorized to moderate various UTAUT relationships.

UTAUT has been applied beyond its first use in organizational contexts to other contexts such as new technology, user populations, consumers, and cultural settings. In addition, new constructs have expanded the range of dependent and independent variables. However, there remains a need for examination of factors that apply to consumer technology use, particularly considering the growth of e-commerce. In the theory, the strongest predictor of technology use and acceptance has been its utilitarian benefits or usefulness. Use of technology for its hedonic or recreational benefits has received scant research attention (Venkatesh et al., 2012). The original UTAUT was revised to integrate three new constructs into UTAUT2, namely (a) intrinsic motivation (hedonic value); (b) price as an explanation for consumer behavior; and (c) habit—a critical factor predicting technology use but not initial acceptance.

Straub, Loch, and Hill (2001) suggested that cultural beliefs were a strong predictor of acceptance of technology. Hofstede's (2001) theory of cultural dimensions was used to explore similarities and differences between US and Chinese students in terms of cultural differences that may affect attitudes toward technology, digital activities, and use of internet shopping features.

Review of the Literature

Use and Acceptance of Technology

The emergence of transportable technology (e.g., small, pocket-size devices such as smartphones and other wireless mobile devices) transformed how individuals perform activities formerly limited to a desktop or laptop computer (Rosen et al., 2013a). Individuals now have 24/7 access to the Internet, e-mail, text messaging, and other applications with a Wi-Fi-enabled mobile device. A national study of 7,446 US smartphone users (18- to 44-year-olds) revealed that eight in ten adults and nine in ten young adults checked their phone within 15 min of awakening (IDC, 2013). Oulasvirta, Rattenbury, Ma, and Raita (2012) found that adults, on average, access their smartphones daily for 34 episodes of short duration (less than 30 s). Other research found that, among US smartphone users, 58% check their phones at least hourly and 73% feel anxious if their phone is misplaced (Mobile Mindset, 2012). Almost half of Japanese student respondents felt insecure, anxious, and perceived they were being ignored when they did not receive a response to their text messages (Kamibeppu & Sugiura, 2005). Rosen et al. (2013b) compared participants with five or more signs of Internet addiction to those with fewer than five signs. Those who were more addicted to the Internet more often used internet searching, shared media, and played video games. Those addicted to the Internet were more anxious about being without technology and more dependent on technology than those not addicted. Anxiety and dependency reflect items on the Internet Addiction Test.

Hedonic and Utilitarian Motivations and Use of Technology

Two main stimuli underlying consumer behavior are hedonic (sensory, experiential) and utilitarian (functional, practical, instrumental) motivations (Babin, Darden, & Griffin, 1994). "Motivation is the energizing force that activates behavior and provides purpose and direction to that behavior" (Hawkins & Mothersbaugh, 2013, p. 352). Motivation can come from physical, emotional, cognitive, or social stimuli that activate and direct behavior. A utilitarian perspective emphasizes efficiency and sensible disbursement of resources, whereas a hedonic perspective emphasizes the shopping experience itself as enjoyable and entertaining (Childers, Carr, Peck, & Carson, 2001). The theory of consumer motivation posits that to satisfy hedonic motives, subjective, psychological information will be used to make decisions but to satisfy utilitarian motives, objective, tangible information will be used (Babin et al., 1994; Hirschman & Stern, 1999; Holbrook & Hirschman, 1982). Hirschman and Stern (1999) found that positive emotions such as pleasure and enjoyment often accompanied experiential shopping and hedonic consumption. Venkatesh et al. (2012, p. 161) defined hedonic motivation related to technology use as the "fun or pleasure derived from using a technology." Hedonic motivation plays an important role in influencing use of consumer technology services (Brown & Venkatesh, 2005; Childers et al., 2001), for example use of recreational shopping features associated with online stores (e.g., e-mail alerts of new products, e-mail alerts of special offers, list of new items, list of best-selling items).

Habit and Use of Technology

"Habit refers to the non-deliberate, automatically inculcated response that individuals may bring to IS usage" (Limayem & Hirt, 2003, p. 66). Habit has been hypothesized to be stored intention (Ajzen, 2002). Fishbein and Ajzen's (1975) theory of reasoned action states that an individual's behavior can be predicted by an intention to behave in a particular way. Further, the theory holds that intentions are shaped by attitudes toward the behavior and by social norms which, in turn, are based on relevant beliefs regarding the behavior in question (Limayem & Hirt, 2003). Habits reveal automatic behavior established during the past, such that a particular cue calls forth the behavior even when an individual does not consciously intend to react. Habit is an important variable in explaining and predicting technology use (Limayem & Hirt, 2003; Limayem, Hirt, & Cheung, 2007).

"Recent work has challenged the role of behavioral intention as the key predictor of technology use and introduced a new theoretical construct (i.e., habit) as another critical predictor of technology use" (Venkatesh et al., 2012, p. 158). Habit has been operationalized as prior behavior (Kim & Malhotra, 2005) and as the extent to which an individual believes the behavior to be automatic (Limayem et al., 2007).

Gender and Use of Technology

The selectivity model (Meyers-Levy, 1989; Meyers-Levy & Maheswaran, 1991; Meyers-Levy & Sternthal, 1991) asserts that women are comprehensive information processors. Women not only react to subtle cues and exhibit greater sensitivity to relevant information but evaluate subjective and objective product attributes in an attempt to assimilate all available cues. In contrast, men are selective information processors; that is, men ignore subtle cues in an attempt to be efficient. Men use prior experience or informal methods such as trial and error repetition relying on readily available salient cues. If the selectivity model is correct, then men may be less likely to use recreational shopping features than women and women may use all internet shopping features more frequently than men.

Gender may be a key variable in acceptance and use of technological services. If gender processing differences do exist, it is important for Web site developers to understand the differences in order to produce shopping features that will be used. A tendency for women to notice subtle cues becomes important when these cues contain relevant information. For example, the implicit level of risk (e.g., financial, product quality) associated with a product could serve as a subtle information cue which men may ignore and women notice.

Computer anxiety has been defined as "aversion, fear or apprehension towards interacting with computers or thinking about computers, intimidation by, resistance to, hostility, or aggression towards computers" (Beckers & Schmidt, 2001, p. 35). Women displayed higher levels of computer anxiety than men (e.g., Bozionelos, 2001; Igbaria & Chakrabarti, 1990). Attitudes (e.g., computer anxiety) are determinants of intention and subsequent behavior.

Perceived ease of use has been an obstacle to user acceptance (Venkatesh & Davis, 2000). Ease of use was more relevant to women in making decisions regarding technology use but usefulness was more relevant to men; ease of use was not important to men at any point in decision making (Venkatesh & Morris, 2000). Rosen et al. (2013a) found that men (vs. women) more often played video games, had more online friends, did more media sharing, and experienced more anxiety about not having access to technology.

Kimbrough, Guadagno, Muscanell, and Dill (2013) used an online survey to collect data about experiences with multiple forms of technology-mediated communication, that is, social networking sites, e-mail, video calls, instant messaging, texting, and phone calls. They found that women, compared to men, preferred and more frequently used technology-mediated communication such as text messaging, social media, and online video calls.

Given the extensive role of technology in e-commerce and the increasing presence of women in online shopping domains, understanding gender differences in technology use decisions is an important issue with economic ramifications for online stores. Gender differences may be more pronounced for some product types (e.g., clothing compared to books; identity-expressive compared to functional products).

Culture and Use of Technology

Loch, Straub, and Kamel (2003) applied the model of cultural influence to look at culture-specific facilitators and barriers to adoption and use of the internet in the Arab world. Both social norms and extent of technological acculturation affected individual acceptance and use of the internet (Straub, Keil, & Brenner, 1997).

Hofstede's (2001) theory of cultural dimensions helps explain how cultural differences may influence technology acceptance and use (Itim International, 2017). Hofstede's six cultural dimensions are individualism (China—20; US—91), indulgence (China—24, US—68), long-term orientation (China—87; US—26), masculinity (China—66; US—62), power distance (China—80; US—40), and uncertainty avoidance (China—30; US—46).

The power distance dimension (PDI) reveals attitudes toward inequality among members of a culture (Itim International, 2017). PDI is defined by Hofstede (2001) as the extent to which less powerful members within a country expect and accept unequal distribution of power. China at 80/100 ranks high in PDI, meaning that inequalities among people are expected and considered acceptable. Formal authority and sanctions influence people's behavior. The USA at 40/100 ranks low in PDI, meaning that citizens accept the uniqueness of each person and, therefore, the inequality, but not the idea of wielding power over others to influence their ideas and behavior.

Individualism (IND) refers to the degree of interdependence among society's members (Hofstede, 2001; Hofstede, Hofstede, & Minkov, 2010; Itim International, 2017). In collectivist (COL) societies, "in-groups" take care of members in exchange for loyalty. In individualist societies, people only look after themselves and their direct family. Self-image is defined in terms of "I" (IND) or "We" (COL). China at 20/100 is a highly collectivist culture where interests of the group are more important than individual interests. The USA at 91/100 is a highly individualist culture as shown by an emphasis on equal rights for all citizens.

A high score on the Masculine (MAS) dimension indicates a society driven by competition, achievement, and success, whereas a low score (Feminine) indicates a society that values caring for others and quality of life (Hofstede, 2001; Hofstede et al., 2010; Itim International, 2017). In a Feminine society, quality of life is a sign of success and standing out from the crowd is disgraceful. China at 66/100 is a Masculine society—competitive and success-oriented to the point that citizens will sacrifice family and leisure priorities to work. Chinese students are concerned about their exam scores and ranking as these criteria are the means to achieve success. The USA at 62/100 is also a Masculine society, sharing values that individuals should "strive to be the best they can be" and a tendency to display and talk freely about their achievements.

Uncertainty Avoidance (UA) refers to the extent to which members of a culture create beliefs and institutions to reduce anxiety and perceived threats caused by ambiguous or unknown situations (Hofstede, 2001; Hofstede et al., 2010; Itim International, 2017). China at 30/100 has a low score on Uncertainty Avoidance meaning that Chinese citizens are adaptable, entrepreneurial, and comfortable with ambiguity. The USA at 46 also has a low score on UA, meaning there is a willingness to try something new or different, acceptance of new ideas and innovative products, and tolerance for freedom of expression.

Long-Term Orientation (LTO) describes a society's links with its past while coping with present and future challenges (Hofstede, 2001; Hofstede et al., 2010; Itim International, 2017). Low scores on LTO indicate normative societies that maintain time-honored traditions and norms and view societal change with suspicion. High scores indicate cultures that take a pragmatic approach and encourage thrift and education as ways to prepare for the future. China at 87/100 scores high on LTO indicating that Chinese adapt traditions to changed conditions, save and invest, practice thriftiness, and persevere in accomplishing results. The USA at 26/100 is a normative society whose citizens have strong opinions about social and governmental issues. Performance is evaluated on a short-term basis, leading to an emphasis on quick results.

Indulgence is defined as the extent to which individuals are socialized to control their desires and impulses (Hofstede, 2001; Hofstede et al., 2010; Itim International, 2017): indulgent (relatively weak control) or restrained (relatively strong control). The relatively weak control of desires and impulses may lead to an enhanced vulnerability to addictive behaviors. China at 24/100 is a restrained society with a tendency to cynicism and pessimism, delayed gratification, and de-emphasis on leisure time. Social norms restrain their behavior, and indulging themselves is considered wrong. The USA at 68/100 is an indulgent society with an interest in immediate gratification, citizens who work hard and play hard, and a high rate of addictive behaviors.

Compared with the USA, China ranks much lower in individualism and indulgence, much higher in long-term orientation and power distance, about the same in masculinity, and lower in uncertainty avoidance. Because members of both cultures are competitive and success-oriented, it is unlikely the Masculine dimension will account for any differences. The two cultures differ greatly in individualism and indulgence suggesting that Chinese students may have less frequent use of digital activities and internet shopping features because these activities are individual and indulgent. US students may lack self-control with regard to digital activities resulting in anxiety about losing access to technology. The two cultures also differed greatly in power distance (inequalities among people) suggesting that gender differences in use of technology may exist or that there may be interactions between culture and gender on technology use. The two cultures also differed greatly in Long-Term Orientation (past, present, and future concerns) suggesting that strong opinions among US citizens about social issues may lead to more negative attitudes toward technology. China scored lower on Uncertainly Avoidance (they are more willing to try something new or different) than the USA, suggesting they may have slightly more positive, and less negative, attitudes toward technology. How individuals obtain, consolidate, and use information depends on how they have learned to process information (de Mooij & Hofstede, 2011). Members of collectivistic cultures such as China are used to symbols, signs, and tacit communication and have learned a different way of processing information than members of individualistic cultures, who have adapted to verbal communication, explanations, and speech (de Mooij & Hofstede, 2011, p. 187). Therefore, because of these cultural differences, attitudes toward technology, habitual use of digital activities, and use of internet shopping features may differ among a sample of male and female Chinese and US college student consumers.

Purpose of the Study and Hypotheses

The purpose of this study was to examine how attitudes toward technology (ATT), habitual use of digital activities, and frequency of use of internet shopping features differ among a sample of male and female Chinese and US college student consumers. The current research measured attitudes toward technology, habit, and both utilitarian and hedonic values. The following variables were held constant in this research: price (use of internet shopping features is provided free of charge to Web site users); voluntariness (use of internet shopping features is voluntary); and age (the college student sample had little variance in age). Understanding the conditions under which convenience and recreational shopping features are or are not used is an important issue given the investment of companies in e-commerce and the effect these shopping features may have on a consumer's decision to purchase products online.

Research has documented the importance of ATT to consumer behavior, but has not compared ATT, habitual use of digital activities, and use of internet shopping features in combination with gender and culture. The following hypotheses were proposed:

- H1a–e: Male and female college students will differ in (a) positive ATT, (b) negative ATT, (c) anxious ATT, (d) use of recreational internet shopping features, and (e) use of convenience internet shopping features.
- H2a–e: Chinese and US college students will differ in (a) positive ATT, (b) negative ATT, (c) anxious ATT, (d) use of recreational internet shopping features, and (e) use of convenience internet shopping features.
- H3a–e: Male and female college students will differ in habitual use of (a) total digital activities, (b) social media, (c) smartphone, (d) e-mail, and (e) entertainment.
- H4a–e: Chinese and US college students will differ in habitual use of (a) total digital activities, (b) social media, (c) smartphone, (d) e-mail, and (e) entertainment.

Method

A questionnaire was compiled that contained demographic items and scales measuring internet shopping features (Kim & LaRose, 2004), ATT, and habitual use of digital activities (Rosen et al., 2013a). The questionnaire was translated into Chinese by the widely used back-translation method.

Instruments. ATT includes items that measure positive attitudes (six items), negative attitudes (three items), anxious and dependent attitudes (three items; Rosen et al., 2013a). A five-point scale (5 = strongly agree; 1 = strongly disagree) measured level of agreement with each item. Following are example items measuring positive attitudes ("I feel it is important to be able to find any information whenever I want online"), anxious attitudes ("I get anxious when I don't have my cell phone"), and negative attitudes ("New technology makes people more isolated"). The scales measure attitudes toward technology in general rather than any specific technology. The items in anxious ATT assess anxiety related to dependence or addiction to technology; not anxiety caused by aversion or fear of computers. Items in each subscale were summed to create a score for that subscale.

Habit was operationally defined as frequency of use. The original scale for measuring use of digital activities (Rosen et al., 2013a) included 44 items. In the interest of shortening the scale, representative items were selected: e-mail (two items), smartphone (four items), entertainment (three items), and social media (two items). Frequency of using digital activities used a 10-point scale (1 = never; 2 = once a month; 3 = several times a month; 4 = once a week; 5 = several times a week; 6 = once a day; 7 = several times a day; 8 = once an hour; 9 = several times an hour; 10 = all the time). Items were summed to create a score for each digital activity and an overall score; higher scores indicate more habitual use and greater addiction to these digital activities. Participants indicated how often they do each of the following activities: Check personal e-mail; check work or school e-mail; send, receive, and check text messages on a mobile phone (for personal use); send, receive, and check text messages on a mobile phone (during work or class); browse the Web on a mobile phone (for personal use); browse the Web on a mobile phone (during work or class); watch TV shows, movies, other media on a TV set; search the internet for information on any device; play games on a computer, video game console, or smartphone; general social media usage (check Facebook page or other social networks at work or school; check Facebook page or other social networks outside work or school).

Internet shopping features include both convenience (five items) and recreational (four items) features (Kim & LaRose, 2004). Frequency of use of internet shopping features was measured with a five-point scale (1 = never; 2 = rarely; 3 = sometimes; 4 = often; 5 = very often) indicating how often these features are used when shopping online. Items measuring use of convenience features are: search engine for finding products I want; product price information; price comparison with other Web sites; product-related articles or reviews written by experts; reviews written by other shoppers. Items measuring use of recreational shopping features are: e-mail alerts of new products; e-mail alerts of special offers; list of new items, and list of best-selling items. Items in each subscale were summed to create a score for that scale.

Procedure. The research was approved by the Institutional Review Board at the US university. The questionnaire was distributed and collected in large lecture classes and took about 20 min to complete. Participants in this study were students from a university in the Midwestern United States and from Lanzhou University in China. In the USA, in 2014, there were 21.6 million college students with an estimated $163 billion in discretionary income (US College Explorer, 2014). In 2015, in China, there were about 26.25 million college students (Number of students …, 2017). Eight million students will graduate from Chinese higher education institutions in 2017, ten times higher than in 1997, and more than twice the number graduating in the USA (Montgomery, 2017; Stapleton, 2017). Therefore, US and Chinese male and female college students were considered an important sample for an investigation of attitudes toward technology, habitual use of digital activities, and use of internet shopping features.

Results

Participants were 209 US and 193 Chinese college students—185 women; 215 men; mean age = 21.13; range 17–32—from a variety of majors (e.g., engineering, sports management, fashion, architecture, business). In the US sample, there were 144 (68.9%) Caucasian, 38 (18.2%) African-American, 12 (5.74%) Hispanic/Latino, 11 (5.25%) Asian/Asian-American, and 4 (2.0%) otherwise classified.

Descriptive statistics and Cronbach's alpha reliability for attitudes toward technology, internet shopping features, and digital activities are displayed in Table 1. Reliability was acceptable for all scales for both countries; Cronbach's alpha ranged from 0.67 to 0.88.

MANOVA, with country and gender as independent variables and positive ATT, negative ATT, anxious/dependent ATT, use of convenience and recreational shopping features as dependent variables, was significant for country [$F(5, 391) = 8.33$, $p < 0.000$], gender [$F(5, 391) = 6.19$, $p < 0.000$], and the country/gender interaction [$F(5, 391) = 5.40$, $p < 0.000$]. ANOVA (see Table 2) revealed that men and women did not differ in positive ATT or negative ATT. Men and women differed in anxious/dependent ATT, use of recreational and convenience shopping features. Women indicated greater anxiety/dependence on technology and more often used both convenience and recreational shopping features than men.

The interaction between country and gender revealed that Chinese men and Chinese women did not differ in use of recreational shopping features but US women, compared with US men, more often used recreational shopping features. US men expressed less anxiety/dependence on technology than v women but Chinese men and women did not differ in anxiety/dependence about technology. H1a and H1b were not supported; H1c, H1d, and H1e were supported.

Table 1 Descriptive statistics and reliability for measures

	Participants	Mean (SD)	Observed range	Reliability Cronbach's α
ATT-positive	All	23.97 (3.75)	11–32	0.77
	US	24.13 (4.23)	12–32	0.83
	CH	23.77 (3.14)	11–32	0.67
ATT-anxious	All	10.16 (2.97)	3–15	0.80
	US	9.50 (3.37)	3–15	0.83
	CH	10.85 (2.28)	3–15	0.70
ATT-negative	All	9.72 (2.72)	3–15	0.76
	US	10.06 (2.97)	3–15	0.78
	CH	9.35 (2.40)	3–15	0.74
Use of convenience shopping features	All	18.98 (3.85)	5–38	0.79
	US	19.21 (4.36)	5–38	0.84
	CH	18.73 (3.21)	5–25	0.71
Use of recreational shopping features	All	11.53 (4.02)	4–20	0.84
	US	11.62 (4.62)	4–20	0.88
	CH	11.46 (3.26)	4–19	0.75
Digital activities	All	67.43 (17.76)	19–110	0.85
	US	77.31 (15.97)	22–110	0.84
	CH	56.74 (12.77)	19–96	0.71

ANOVA revealed that Chinese and US students did not differ in positive ATT, use of convenience or recreational shopping features. Chinese and US students did differ in negative ATT and in anxious/dependent ATT. Chinese students indicated higher anxiety/dependence on technology than US students; US students indicated more negative attitudes toward technology than Chinese students. H2a, H2d, and H2e were not supported; H2b and H2c were supported.

A second MANOVA, with country and gender as independent variables and habitual use of digital activities as dependent variables (total digital activities, social media, smartphone, e-mail, and entertainment), was significant for country [$F(4, 394) = 108.93$, $p < 0.000$] and gender [$F(4, 394) = 7.60$, $p < 0.000$] but the country/gender interaction was not significant [$F(4, 394) = 1.81$, $p < 0.126$]. ANOVA showed that men and women differed in habitual use of total digital activities, social media, and smartphone, but did not differ in habitual use of e-mail or entertainment (see Table 3). Women engaged in total digital activities, social media, and smartphone to a greater degree than men. H3a, H3b, and H3c were supported; H3d and H3e were not supported.

Chinese and US students differed in habitual use of total digital activities, social media, smartphone, e-mail, and entertainment (see Table 3). US students indicated more habitual use of all digital activities than Chinese students. H4a, H4b, H4c, H4d, and H4e were supported.

Table 2 ANOVA results of country and gender for scores on attitudes toward technology and use of Internet shopping features (only significant interactions are shown)

Scale		Mean (SD)	d.f.	MS	*F*	*p*<
ATT-positive						
Country			1, 398	13.34	0.955	0.329
	US (*n* = 206)	24.11 (4.23)				
	CH (*n* = 193)	23.77 (3.14)				
Gender			1, 398	4.95	0.354	0.552
	Men (*n* = 214)	24.05 (3.69)				
	Women (*n* = 185)	23.83 (3.79)				
ATT-anxious						
Country			1, 398	167.64	20.67	0.000
	US (*n* = 206)	9.48 (3.36)				
	CH (*n* = 193)	10.85 (2.78)				
Gender			1, 398	57.50	7.09	0.008
	Men (*n* = 214)	9.76 (2.86)				
	Women (*n* = 185)	10.59 (3.03)				
Country X gender interaction			1, 398	41.58	5.13	0.024
	Men—USA (*n* = 114)	8.85 (3.11)				
	Men—China (*n* = 100)	10.80 (2.12)				
	Women–US (*n* = 92)	10.26 (3.50)				
	Women—China (*n* = 93)	10.91 (2.45)				
ATT-negative						
Country			1, 398	49.79	6.77	0.010
	US (*n* = 206)	10.07 (2.97)				
	CH (*n* = 193)	9.35 (2.40)				
Gender			1, 398	4.35	0.59	0.442
	Men (*n* = 214)	9.64 (2.91)				
	Women (*n* = 185)	9.82 (2.50)				
Use of Internet shopping features—convenience						
Country			1, 398	21.88	1.48	0.224
	US (*n* = 206)	19.18 (4.37)				
	CH (*n* = 193)	18.73 (3.20)				
Gender			1, 398	45.79	3.10	0.079
	Men (*n* = 214)	18.66 (3.88)				
	Women (*n* = 185)	19.32 (3.79)				

(continued)

Table 2 (continued)

Scale		Mean (SD)	d.f.	MS	*F*	*p*<
Use of Internet shopping features—recreational						
Country			1, 398	14.53	.99	0.321
	US (n = 206)	11.66 (4.63				
	CH (n = 193)	11.45 (3.26)				
Gender			1, 398	314.84	21.38	0.000
	Men (n = 214)	10.71 (3.69)				
	Women (n = 185)	12.53 (4.17)				
Country X gender interaction			1, 398	282.06	19.15	0.000
	Men—USA (n = 114)	10.11 (3.89)				
	Men—China (n = 100)	11.41 (3.33)				
	Women—USA (n = 92)	13.58 (4.77)				
	Women—China (n = 93)	11.51 (3.19)				

Table 3 ANOVA results of country and gender on digital activities

Scale	Mean (SD)	d.f.	Mean square	*F*	*p*<
Digital activities—total					
Country		1, 397	43006.12	205.72	0.000
USA	77.31 (15.97)				
CH	56.74 (12.77)				
Gender		1, 397	738.43	3.53	0.061
Men	66.45 (17.33)				
Women	68.51 (18.27)				
Digital activities—social media					
Country		1, 397	1523.56	64.61	0.000
USA	13.63 (4.89)				
CH	9.79 (5.07)				
Gender		1, 397	475.31	20.16	0.000
Men	10.83 (5.40)				
Women	12.88 (5.01)				
Digital activities—smartphone					
Country		1, 397	3363.40	77.22	0.000
USA	30.57 (6.68)				
CH	24.84 (6.56)				

(continued)

Table 3 (continued)

Scale	Mean (SD)	d.f.	Mean square	*F*	*p*<
Gender		1, 397	175.70	4.03	0.045
Men	27.28 (7.29)				
Women	28.42 (7.08)				
Digital activities—e-mail					
Country		1, 397	5430.13	427.03	0.000
USA	14.13 (3.94)				
CH	6.81 (3.17)				
Gender		1, 397	0.006	0.000	0.983
Men	10.71 (4.71)				
Women	10.49 (5.57)				
Digital activities—entertainment					
Country		1, 397	1344.11	45.74	0.000
USA	18.98 (5.60)				
CH	15.30 (5.22)				
Gender		1, 397	63.36	2.16	0.000
Men	17.63 (5.51)				
Women	16.72 (5.94)				

Discussion and Implications

Summary of Results and Discussion

Culture and gender influenced more habitual use of digital activities, positive, negative, and anxious/dependent ATT, and use of internet shopping features. US and Chinese students shared some attitudes and behaviors (i.e., positive ATT, use of convenient and recreational internet shopping features) but differed in some attitudes and behaviors (i.e., negative and anxious/dependent ATT, habitual use of digital activities). US students indicated more habitual use of all digital activities than Chinese students. Men and women did not differ in positive or negative ATT but women had greater anxiety/dependence on technology than men did. Women indicated more habitual use of total digital activities, social media, and smartphone than men.

This research focused on a consumer context and examined relevant variables that provide an increased understanding of acceptance and use of technology. Habit as operationalized in this study was measured as frequency of use of digital activities. US students indicated more habitual use as might be expected in an indulgent society, where there are weak controls over indulgence of desires and impulses. But Chinese students indicated higher anxiety/dependence on technology ("I am dependent on my technology") than US students which was an unexpected result since China is considered to be a restrained country. However, the results are consistent with Ni, Yan, Chen, and Liu (2009) who surveyed first-year students at a university in China and found that more than 6% were addicted to the Internet. Internet-addicted students (vs. non-internet addicted) had higher scores on anxiety; internet addiction was positively correlated with anxiety. Men (6.6%) were slightly more likely to be internet addicted than women (6.1%).

US women indicated more habitual use of total digital activities, social media, and smartphone than men and indicated greater anxiety/dependence on technology than men. These findings may reflect the results of socialization in an indulgent society where there is increased vulnerability to addictive behavior, especially for women. Women, compared to men, have been found to engage in several types of addictive behavior such as compulsive shopping (Lee & Workman, 2015), eating disorders (Cook et al., 2014; Woodside et al., 2001; Yates, 2013), and other addictive disorders (Cook et al., 2014; Fattore, Melis, Fadda, & Fratta, 2014; Yates, 2013).

US students indicated more negative attitudes toward technology than Chinese students which may be partially explained by differences in Long-Term Orientation. The USA had low scores on LTO indicating a normative culture with suspicious attitudes toward societal changes and expression of strong opinions about social issues such as addressed in the measure of negative attitudes toward technology (e.g., new technology makes people waste too much time, makes life more complicated, and makes people more isolated). However, China's high score on LTO plus a low score on Uncertainty Avoidance indicates that Chinese students are likely to be receptive to, and adapt readily to, changes in technology.

US and Chinese students did not differ in use of convenience and recreational internet shopping features. Chinese men and women did not differ in use of convenience and recreational internet shopping features. Information gathering is often the first step in a purchase decision. "Whereas in individualistic cultures of low power distance, people will actively acquire information via the media and friends to prepare for purchases, in collectivistic and/or high power distance cultures, people will acquire information more via implicit, interpersonal communication and base their buying decisions on feelings and trust in the company"(de Mooij & Hofstede, 2011, p. 188). Zheng, Favier, Huang, and Coat (2012) concluded that Chinese consumers reduced risk by searching for information about the product and price or by visiting a store to look at the product. Online reviews were not preferred as a means of risk reduction because consumers either enjoyed searching for information on the internet or believed online reviews were not trustworthy. Consistent with Zheng et al.'s study, in the current study, Chinese men and women used both convenience and recreational internet shopping features to search for information.

US women more frequently used both convenience and recreational shopping features than men did. The selectivity model (Meyers-Levy, 1989; Meyers-Levy & Maheswaran, 1991; Meyers-Levy & Sternthal, 1991) may partially explain these findings; that is, the model asserts that women assimilate all available cues containing relevant information but men rely on readily available salient cues and use prior experience or informal methods such as trial and error to make purchase decisions. The selectivity model was developed in the USA and may not explain differences in use of internet shopping features among men and women in other cultures.

Theoretical Implications

Venkatesh et al. (2012) observed that use of technology for its hedonic or recreational benefits has not received much research attention. Therefore, they revised the original Unified Theory of Acceptance and Use of Technology to integrate three new constructs: (a) intrinsic motivation (hedonic value); (b) price as an explanation for consumer behavior; and (c) habit—a factor predicting technology use but not initial acceptance. The current study included measures of intrinsic motivation and habit (but not price) and found support for these variables as influences on technology use between two cultures and between men and women.

Based on Hofstede's (2001) theory of cultural dimensions, differences were predicted among a sample of male and female Chinese and US consumers regarding attitudes toward technology, habitual use of digital activities, and frequency of use of internet shopping features. Partial or full support was found for all four major hypotheses formulated based on cultural differences in the values of individualism, indulgence, long-term orientation, power distance, and uncertainty avoidance. Results confirmed the influence of culture on attitudes toward technology, habitual use of digital activities, and frequency of use of internet shopping features. No previous research has explored the relationships among these three variables between participants from the USA and China. These findings can contribute to the body of research using Hofstede's theory to investigate cultural differences.

Implications for Chinese Consumers and the Fashion Market

Results of this study revealed that Chinese male and female college students did not differ in their anxiety/dependence on technology, use of internet shopping features, and habitual use of total digital activities, social media, and smartphone. These results can provide useful information for e-retailers or virtual store managers and developers as they plan selective marketing strategies to serve their young target market in China. For example, Chinese men and women used both convenience and recreational internet shopping features to search for information. Because Chinese consumers are a key group in a global fashion market, their use of internet shopping features in online shopping malls will greatly impact growth of the online fashion market. What additional internet shopping features would appeal to Chinese consumers? Perhaps a consumer survey would produce an answer to that question.

In the current study, Chinese students indicated some degree of addiction to digital activities as measured by frequency of use. Based on a 2015 iResearch report, the four top reasons Chinese consumers engaged in online shopping were quality assurance, price, brand preference, and unavailability of product locally (iResearch, 2015). Popular categories that reveal an attraction to luxury and fashion items include clothes, shoes, accessories, perfume, jewelry, and watches—categories that may be hard to find locally and may be cheaper online (Fu, 2014; Rapp, 2016). The growing desire for foreign luxury and fashion items has potential to lead to addictive internet shopping behaviors.

Because Chinese score is low on uncertainty avoidance, indicating a willingness to try something new or innovative, the Chinese fashion market might be a good testing ground for design and deployment challenges associated with new types of wearable and immersive technology (Saleem et al., 2017). Chinese consumers may favorably receive an expansion of the online shopping network to digital fashion as a space for creation, access, and social sharing.

Public Policy Implications

Given the increasing availability and affordability of transportable technology (e.g., smartphones and other wireless mobile devices), technology addiction may become more prevalent. Digital over-engagement can have negative effects on daily lives including compulsive buying and diminishing participation in civic activities (Montag & Walla, 2016). Public policy intervention can address how digital addiction and online shopping behaviors are related.

There are policy issues regarding the digital divide across regions, education, income, generations, and gender that influence accessibility to digital tools and activities. How can the digital gap be reduced? Public policy is required to explore where the digital divide of technology acceptance and use among different groups originates. Internet exclusion is closely related to marketing issues regarding who can or cannot conveniently participate in online shopping behaviors. Further, public policy issues related to privacy, freedom of expression, and human rights may be raised in regard to digital activities of marketing.

Limitations and Implications for Further Research

This study was limited to data from Chinese and US college students. Lanzhou University in China is located in Gansu Province which is an inland location and, thus, not a representative area of China for examining technology use. College students, in general, cannot be considered representative of the general population. College students are an important target market for many businesses; however, a general population survey would provide a more complete picture of variables affecting technology acceptance and use. Another limitation is the scale, published in 2004, used to measure frequency of use of internet shopping features; undoubtedly both convenience and recreational internet shopping features have been added to internet shopping sites in the last 13 years, for example links to fashion bloggers. To shorten the original 44-item scale measuring use of digital activities, 11 representative items were selected. Shortening the scale may have affected its effectiveness in measuring habitual use of digital activities.

Another limitation relates to external factors present in different cultures such as a country's particular situation, government regulations, or business competition that could influence consumers' perception about adopting technology. This study did not include these factors so further study can be examined with these external issues. Also, we did not measure if the variables used in the study (e.g., attitudes toward technology, Internet shopping features, digital activities) affect consumers' purchasing intention or behavior regarding fashion items. Further research is recommended to expand this study to other countries examining cultural differences in technology acceptance and use.

There are three promising research areas regarding digital technology and cultural differences in internet shopping. First, there are research questions about how attitude toward digital tools and activities influences internet shopping behavior and how these attitudes vary across developed and developing countries. Second, longitudinal data may help explain how individual preferences and digital activities during childhood influence consumer behavior in both offline and online venues during adulthood. Third, wearable technology provides a continuous connection between business and consumer but when does this connection become an annoyance or an intrusion into personal lives?

References

Ajzen, I. (2002). Residual effects of past on later behavior: Habituation and reasoned action perspectives. *Personality & Social Psychology Review, 6*(2), 107–122.

Babin, B. J., Darden, W. R., & Griffin, M. (1994). Work and/or fun: Measuring hedonic and utilitarian shopping value. *Journal of Consumer Research, 20*(4), 644–656.

Beckers, J. J., & Schmidt, H. G. (2001). The structure of computer anxiety: A six-factor model. *Computers in Human Behavior, 17*(1), 35–49.

Bose, S. (2016, December 6). 8 in 10 US internet users now buy online, says GlobalWebIndex. Retrieved from https://smallbiztrends.com/2016/12/2016-online-shopping-statistics.html.

Bozionelos, N. (2001). Computer anxiety: Relationship with computer experience and prevalence. *Computers in Human Behavior, 17*(2), 213–224.

Brown, S. A., & Venkatesh, V. (2005). Model of adoption of technology in the household: A baseline model test and extension incorporating household life cycle. *MIS Quarterly, 29*(4), 399–426.

Childers, T. L., Carr, C. L., Peck, J., & Carson, S. (2001). Hedonic and utilitarian motivations for online retail shopping behavior. *Journal of Retailing, 77*, 511–535.

Cook, B., Engel, S., Crosby, R., Hausenblas, H., Wonderlich, S., & Mitchell, J. (2014). Pathological motivations for exercise and eating disorder specific health-related quality of life. *International Journal of Eating Disorders, 47*(3), 268–272.

de Mooij, M., & Hofstede, G. (2011). Cross-cultural consumer behavior: A review of research findings. *Journal of International Consumer Marketing, 23,* 181–192.

E-commerce drives retail sales growth in China. (2015, September 25). eMarketer.

E-commerce in China: Driving a new consumer culture. (2014, January). KPMG Global China Practice.

Fattore, L., Melis, M., Fadda, P., & Fratta, W. (2014). Sex differences in addictive disorders. *Frontiers in Neuroendocrinology, 35*(3), 272–284.

Fishbein, M., & Ajzen, I. (1975). *Belief, attitude, intention and behavior.* Reading, MA: Addison-Wesley.

Fu, R. (2014, September 4). Chinese cross-border online Shoppers insights. *China Internet Watch.*

Hawkins, D., & Mothersbaugh, D. (2013). *Consumer behavior: Building marketing strategy* (12th ed.). New York: McGraw Hill/Irwin.

Hirschman, E. C., & Stern, B. B. (1999). The roles of emotion in consumer research. *Advances in Consumer Research, 26,* 4–11.

Hofstede, G. (2001). *Culture's consequences: Comparing values, behaviors, institutions and organizations across nations.* Newbury Park, CA: Sage.

Hofstede, G., Hofstede, G. J., & Minkov, M. (2010). *Cultures and organizations: Software of the mind* (Revised and Expanded 3rd ed.). New York: McGraw-Hill.

Holbrook, M. B., & Hirschman, E. C. (1982). The experiential aspects of consumption: Consumer fantasies, feelings, and fun. *Journal of Consumer Research, 9*(2), 132–140.

IDC. (2013). Always connected: How smartphones and social media keep us engaged. IDC Research. Retrieved http://www.scribd.com/doc/133393152/IDC-Facebook-Always-Connected.

Igbaria, M., & Chakrabarti, A. (1990). Computer anxiety and attitudes towards microcomputer use. *Behaviour & Information Technology, 9*(3), 229–241.

Internet Trends. (2017). Stats & Facts in the US and Worldwide. Retrieved from https://www.vpnmentor.com/blog/vital-internet-trends/.

Internet Users by Country. (2016). Internet Live Stats. Retrieved from www.internetlivestats.com/internet-users-by-country/.

iResearch. (2015). 2015 China cross-border online shoppers report. Retrieved from www.iresearchchina.com/content/details8_19346.html.

Itim International. (2017). Geert Hofstede cultural dimensions. Retrieved from http://www.geert-hofstede.com.

Kamibeppu, K., & Sugiura, H. (2005). Impact of the mobile phone on junior high-school students' friendships in the Tokyo metropolitan area. Cyberpsychology & Behavior, 8(2), 121–130.

Kim, J., & LaRose, R. (2004). Interactive e-commerce: Promoting consumer efficiency or impulsivity? *Journal of Computer-Mediated Communication, 10*(1), 00–00.

Kim, S. S., & Malhotra, N. K. (2005). A longitudinal model of continued IS use: An integrative view of four mechanisms underlying post-adoption phenomena. *Management Science, 51*(5), 741–755.

Kimbrough, A. M., Guadagno, R. E., Muscanell, N. L., & Dill, J. (2013). Gender differences in mediated communication: Women connect more than do men. *Computers in Human Behavior, 29,* 896–900.

Lee, H., Gillai, B., Chen, Y., & Rammohan, S. (2016). US-to-China B2C E-commerce: Improving logistics to grow trade. *Stanford Value Chain Innovation Initiative White Paper.* Retrieved from https://www.gsb.stanford.edu/sites/gsb/files/publication-pdf/vcii-white-paper-us-china-ecommerce-imrpove-logistics-grow-trade.pdf.

Lee, S.-H., & Workman, J. (2015). Compulsive buying and branding phenomena. *Journal of Open Innovation: Technology, Market, and Complexity, 1,* 3. https://doi.org/10.1186/s40852-015-0004-x.

Limayem, M., & Hirt, S. G. (2003). Force of habit and information systems usage: Theory and initial validation. *Journal of the Association for Information Systems, 4*(1), 65–97.

Limayem, M., Hirt, S. G., & Cheung, C. M. (2007). How habit limits the predictive power of intention: The case of information systems continuance. *MIS Quarterly, 31*(4), 705–737.

Loch, K., Straub, D., & Kamel, S. (2003). Diffusing the Internet in the Arab world: The role of social norms and technological culturation. *IEEE Transactions on Engineering Management, 50,* 45–63.

Martin, C. (2016, February 3). Mobile influence on store sales tops $1 trillion. Retrieved from https://www.mediapost.com/publications/article/268210/mobile-influence-on-store-sales-tops-1-trillion.html.

Meyers-Levy, J. (1989). Gender differences in information processing: A selectivity interpretation. In P. Caferata & A. Tybout (Eds.), *Cognitive and affective responses to advertising* (pp. 219–260). Lexington, MA: Lexington.

Meyers-Levy, J., & Maheswaran, D. (1991). Exploring differences in males' and females' processing strategy. *Journal of Consumer Research, 18*(1), 63–70.

Meyers-Levy, J., & Sternthal, B. (1991). Gender differences in the use of message cues and judgments. *Journal of Marketing Research, 28*(February), 84–96.

Mobile mindset. (2012). Retrieved from https://www.mylookout.com/resources/reports/mobile-mindset

Montag, C., & Walla, P. (2016). Carpe diem instead of losing your social mind: Beyond digital addiction and why we all suffer from digital overuse. *Cogent Psychology, 3*(1). https://doi.org/10.1080/23311908.2016.1157281.

Montgomery, C. (2017, May 26). China's higher education megaproject. *University World News*, 461. Retrieved from http://www.universityworldnews.com/article.php?story=20170523001225119.

Ni, X., Yan, H., Chen, S., & Liu, Z. (2009). Factors influencing internet addiction in a sample of freshmen university students in China. *Cyberpsychology & Behavior, 12*(3), 327–330.

Number of students at universities in China between 2005 and 2015 (in millions). Retrieved from https://www.statista.com/statistics/227028/number-of-students-at-universities-in-china/.

Oulasvirta, A., Rattenbury, T., Ma, L., & Raita, E. (2012). Habits make smartphone use more pervasive. *Personal and Ubiquitous Computing, 16*(1), 105–114.

Rapp, J. (2016, June 22). Revolve CEO: 'The Consumer Came to Us' in China's Fashion E-Commerce Market. *Jing Daily*.

Rosen, L., Whaling, K., Carrier, L., Cheever, N., & Rokkum, J. (2013a). The media and technology usage and attitudes scale: An empirical investigation. *Computers in Human Behavior, 29,* 2501–2511.

Rosen, L. D., Whaling, K., Rab, S., Carrier, L. M., & Cheever, N. A. (2013b). Is Facebook Creating "iDisorders"? The link between clinical symptoms of psychiatric disorders and technology use, attitudes and anxiety. *Computers in Human Behavior, 29,* 1243–1254.

Saleem, K., Shahzad, B., Orgun, M. A., Al-Muhtadi, J., Rodrigues, J. J., & Zakariah, M. (2017). Design and deployment challenges in immersive and wearable technologies. *Behaviour & Information Technology*, 1–12.

Stapleton, K. (2017, April 10). Inside the world's largest higher education boom. *The Conversation*. Retrieved from http://theconversation.com/inside-the-worlds-largest-higher-education-boom-74789.

Statistical Report on Internet Development in China. (2016, July). Retrieved from: http://cnnic.com.cn/IDR/ReportDownloads/201611/P020161114573409551742.pdf.

Sterling, G. (2017, February 17). Reports: Digital, especially mobile, driving trillions in offline retail spending. Retrieved from https://marketingland.com/reports-digital-especially-mobile-driving-trillions-offline-retail-spending-207037.

Straub, D. W., Keil, M., & Brenner, W. (1997). Testing the technology acceptance model across cultures: A three country study. *Information & Management, 33,* 1–11.

Straub, D. W., Loch, K., & Hill, C. (2001). Transfer of information technology to the Arab world: A test of cultural influence modeling. *Journal of Global Information Management, 9,* 6–28.

Tong, F. (2016, Jan 27). China's online retail sales grow a third to $589 billion in 2015. *Internet Retailer*.

US Census Bureau News. (2016, May 17). Retrieved from www.census.gov/retail/mrts/www/data/pdf/ec_current.pdf.

US College Explorer (2014). Retrieved from http://www.refuelagency.com/wp-content/themes/refuelagency/college_explorer/college_explorer_2014_high_level_review.pdf.

Venkatesh, V., & Davis, F. D. (2000). A theoretical extension of the technology acceptance model: Four longitudinal field studies. *Management Science, 46*(2), 186–204.

Venkatesh, V., & Morris, M. G. (2000). Why don't men ever stop to ask for directions? Gender, social influence, and their role in technology acceptance and usage behavior. *MIS Quarterly, 24*(1), 115–139.

Venkatesh, V., Thong, J., & Xu, X. (2012). Consumer acceptance and use of information technology: Extending the unified theory of acceptance and use of technology. *MIS Quarterly, 36*(1), 157–178.

Woodside, D. B., Garfinkel, P. E., Lin, E., Goering, P., Kaplan, A. S., Goldbloom, D. S., et al. (2001). Comparisons of men with full or partial eating disorders, men without eating disorders, and women with eating disorders in the community. *American Journal of Psychiatry, 158*(4), 570–574.

Yates, A. (2013). *Compulsive exercise and the eating disorders: Toward an integrated theory of activity*. London, UK: Routledge.

Zheng, L., Favier, M., Huang, P., & Coat, F. (2012). Chinese consumer perceived risk and risk relievers in e-shopping for clothing. *Journal of Electronic Commerce Research, 13*(3), 255.

Part III
Sustainable Fashion in China

Young Consumers' Perceptions of Sustainable Clothing: Empirical Insights from Chinese Post-90s' College Students

Jin Su, Kittichai (Tu) Watchravesringkan and Jianheng Zhou

Abstract The scale of environmental and social impacts from global apparel production and consumption makes sustainability increasingly important in the contemporary business environment. This study explores the Chinese young consumers' perceptions of and attitudes toward sustainable clothing. Specifically, the study aims to answer the following two research questions: (1) How Chinese college students view and feel about sustainable clothing in terms of their perceived risks, and (2) How their sustainability knowledge, their personal values, and perceived risks impact their attitude toward and willingness to buy sustainable apparel products. The study results based on data collected from 381 Chinese post-90s' college students demonstrate that apparel sustainability knowledge strongly and positively predicts consumer attitude and willingness to buy. The study also reveals that consumer attitude and willingness to buy are significantly predicted by personal values and consumer perceived risks toward sustainable clothing. The study provides valuable baseline information for the Chinese textile and apparel industry and global sustainable textile and apparel marketers.

Keywords Sustainable clothing · Sustainability knowledge · Value Attitude · College student · Millennial generation · China

---

J. Su (✉) · K. (Tu) Watchravesringkan
Department of Consumer, Apparel, and Retail Studies, Bryan School of Business and Economics, The University of North Carolina at Greensboro, Greensboro, NC 27402, USA
e-mail: j_su@uncg.edu

K. (Tu) Watchravesringkan
e-mail: k_watchr@uncg.edu

J. Zhou
Glorious Sun School of Business & Management, Donghua University, Shanghai 200051, China
e-mail: zjh001@dhu.edu.cn

Y. Xu et al. (eds.), *Chinese Consumers and the Fashion Market*, Springer Series in Fashion Business, https://doi.org/10.1007/978-981-10-8429-4_5

Introduction

As the industry supplying and marketing one of the major consumer product categories—clothing, the apparel industry has specific negative impacts on the environment through all stages of the apparel product life cycle, from fiber growth and manufacturing, fabric and garment dyeing and printing, transportation to stores and selling, to disposal at the end of garment life (Hill & Lee, 2012). Some of the key social and environmental sustainability issues in the clothing industry include the increased volume of clothing consumption fueled by "fast fashion"; employee working conditions and wages; pesticide use, excessive chemical usage, and disposal of used garments (Goworek, Fisher, Cooper, Woodward, & Hiller, 2012).

There is an increasing public demand that corporations take responsibility for environmental and societal problems (Dickson, 2000). The literature suggests a consumer-centered approach to sustainability, in which companies present a holistic application of sustainability strategies, well integrated into company policies and product offerings (Hill & Lee, 2012). Textile and apparel firms have been implementing a variety of eco-conscious initiatives throughout their supply chains (Dickson, Loker, & Eckman, 2009; Hiller Connell, 2011). However, as the industry works toward being more environmentally and socially conscious, it is equally important to involve consumers in the process and encourage sustainable consumption of apparel products. Sustainable consumption is also called as ethical consumption, described as making purchase decisions not only on the basis of personal interests but also on the basis of the interests of society and the environment; therefore, it is closely connected with social and environmental sustainability (Goworek et al., 2012). Ethical consumerism has resulted in a market for ethical fashion which is a new approach of "fashion with conscience" (Joergens, 2006).

With the significant achievements of economic development from over 30 years of rapid urbanization and industrialization, China is facing serious environmental deterioration. As China is a major powerhouse of global apparel production, it is undeniable that awareness of environmental and social problems, including pollution, and labor issues and working condition, has grown in recent years in China. Interest in sustainable clothing production and consumption is growing, as evidenced by more and more designers and businesses promote sustainable lifestyle in China (Cerini, March 14, 2016; Song, April 24, 2017). The rapid increase in the size of the sustainable clothing market implies consumers' increasing concern about sustainability. However, most research on sustainable consumption has originated from, and focused on, Western countries; academic research attention to consumer sustainable consumption behavior in China has been very limited so far (Kolk, Hong, & Van Dolen, 2010). In addition, young Chinese consumers' perceptions of and attitudes toward sustainable clothing in China have hardly been studied so far. Thus, this study aims to fill the literature gap by examining Chinese post-90s' college students' perceptions of and consumer behavior toward sustainable clothing.

Previous research consistently indicates that knowledge is a determinant of eco-conscious consumer behaviors (Hill & Lee, 2012; Hiller Connell, 2010; Hwang, Lee, & Diddi, 2015). The literature also reveals the negative effect of perceived risk on consumers' attitudes (Beneke, Flynn, Greig, & Mukaiwa, 2013; Kang & Kim, 2013; Sweeney, Soutar, & Johnson, 1999). Furthermore, the literature maintains that consumers' personal values are associated with the formation of beliefs and attitudes, and often influence intentions to behave in a certain way (Ajzen & Fishbein, 1980). Previous studies explored personal values and socially responsible consumption in the USA and UK (Ma & Lee, 2012). However, little research examined Chinese young consumers' sustainability knowledge, their personal values, and their attitudes toward sustainable consumption. Considering the huge social and cultural impacts that Chinese young generation plays in the Chinese society and their market potential in Chinese apparel retail industry, built upon previous studies, it is crucial to explore the Chinese young consumers' perceptions of and attitudes toward sustainable clothing. Specifically, this study aims to answer the following two research questions: (1) How Chinese college students view and feel about sustainable clothing in terms of their perceived risks, and (2) How their sustainability knowledge, personal values, and perceived risks impact their attitudes toward and willingness to buy sustainable apparel products.

In the following section, the relevant literature is reviewed. Next, the research method employed to investigate the research questions is described. The empirical results of the survey are then presented. The concluding section discusses the implications of the findings, notes the limitations of the study, and offers suggestions for future research.

Literature Review

Chinese Post-90s: Emerging Consumers for Sustainable Apparel Products

Previous research indicates that young consumer segment is a main target consumer group for companies with corporate social responsibility (CSR) attributes, and these young consumers represent a large and powerful consumer segment (Hill & Lee, 2012; Hwang et al., 2015). Young consumers are only now entering their young adulthood; thus, their consumer purchase decisions will impact fashion retail industry for a long time. There are three reasons that motivate us to study Chinese post-90s' college students' perspective. First, given the current vibrant retail environment in China, young consumers are considered one of the most important market segments for apparel products (O'Cass & Choy, 2008; O'Cass & Siahtiri, 2014). Retailers have been attracted to this consumer segment due to its size and consumer spending power. In 2016, the total population in China is 1.37 billion people; however, noticeably the population count in the age group of 17–26 (born between 1990 and 1999) is about 202.1 million people, constituting about 14.7% of China's total population (U.S. Census Bureau, 2017). To put the power of this consumer segment into perspective, the head count of China's post-90s alone is about 62.4% of the entire US population (U.S. Census Bureau, 2017).

Second, the current Chinese post-90s have been growing up with China's rapid economic growth, a fact which affects their daily lives. Most Chinese post-90s are the only child in their family; as such, they represent a generation of young adults whose lifestyle is quite different from the previous generations. They are technologically connected to the global marketplace. They highly regard the power of economic freedom, even social freedom, and are aware of emerging global issues. This generation appreciates sustainable lifestyle and is interested in health and well-being issues in China (Rapp, 2016). Third, over the past decade, the Chinese government has realized the importance of educating young students about sustainability for China's sustainable development (Simões, 2016). Sustainability is heavily promoted in China through all different news and media channels (Kolk et al., 2010; Simões, 2016). As such, understanding Chinese post-90s' perspectives of sustainable clothing offers valuable insights to global sustainable apparel marketers.

Apparel Sustainability Knowledge

McDonald et al. (2009) found consumers process sustainability issues in decision making differently among product categories. General environmental concern may not carry over to the apparel industry specifically (Gam, 2011; Hiller Connell, 2010). Even with an increasing universal awareness of environmental and ethical issues, consumers who purchase apparel are often confused by the meaning of sustainability (Hiller Connell, 2010). Previous studies found that consumers have merely a broad awareness of environmental issues and lack knowledge specific to the apparel industry (Gam & Banning, 2011; Hiller Connell, 2010). Even consumers who are environmentally conscious have little knowledge of the environmental impact of apparel purchases (Hiller Connell, 2010). The notion of ethical clothing is complex as reflected by the use of various terms such as eco, organic, fair-trade, or recycled apparel. In addition, inclusion of sustainability in apparel purchasing decisions may be especially complicated due to additional evaluative criteria such as fit and aesthetic preferences (Gam, 2011; Hiller Connell, 2010).

Hill and Lee (2012) stated that specific understanding of the sustainable issues of the apparel industry is important, as the production and retailing of apparel products represents a large industry with great adverse environmental impacts. Students who were more knowledgeable about textile and apparel environmental issues perceived themselves to have more impact on the problems (Dickson, 2000; Hill & Lee, 2012; Hwang et al., 2015; Hyllegard, Ogle, & Yan, 2014). Additionally, some demographics, including age, gender, and education, have been associated with a variety of environmentally and socially responsible apparel consumption behaviors (Butler

& Francis, 1997; Dickson, 2000; Gam, Cao, Farr, & Kang, 2010; Hill & Lee, 2012). Young consumers hold more favorable attitudes toward environmental regulation and about conserving the environment with sustainable clothing consumption practices (Butler & Francis, 1997; Dickson, 2000; Hill & Lee, 2012; Hyllegard et al., 2014).

Knowledge shapes an individual's beliefs, and attitude is derived from a group of beliefs that one holds about the object of the behavior (Ajzen & Fishbein, 1980). Consumer knowledge of environmental and social issues may play a role in their environmentally and socially responsible apparel consumption as well (Dickson, 2000; Hwang et al., 2015). Knowledge is frequently a prerequisite for consumers to engage in pro-environmental and prosocial behaviors, and a lack of knowledge is a constraint. Consumers with knowledge and concerns regarding environmental and societal issues would be motivated to purchase green and fair-trade products (Hwang et al., 2015). Chen and Burns (2006) maintained that consumer decisions in product purchasing play a vital role in making the textile industry move toward "green" or environmentally responsible production. Previous studies have reported that consumers' higher level of awareness of company's CSR attributes had positive effects on their purchase intentions (Bhaduri & Ha-Brookshire, 2011; Hwang et al., 2015; Smith, 2003). Different levels of consumers' knowledge about sustainability issues in the textile and apparel industry would result in different consumer behaviors, and consumers' apparel sustainability knowledge is the key to help consumers make the right choice. Therefore, based on the previous studies, the following hypothesis was proposed:

H1: There is a positive influence of consumers' apparel sustainability knowledge on (a) consumer attitude toward and (b) willingness to buy sustainable clothing.

Perceived Risks of Purchasing Sustainable Clothing

The issues surrounding consumers' decisions to try new brands and adopt innovative products are of both theoretical interest and pragmatic importance. One factor that research has identified as a critical determinant of consumers' willingness to buy a new item is the perceived risks associated with the purchase (Grewal, Gotlieb, & Marmorstein, 1994). Risk is a consumer's perception of the uncertainty and adverse consequences of engaging in an activity (Chang & Tseng, 2013; Dowling & Staelin, 1994). Perceived risk is commonly thought of as felt uncertainty regarding possible negative consequences of using a product or service. Dowling (1986) defined perceived risk as the uncertainty of a desired performance that customers experience when making purchasing decisions. When consumers perceive higher risks, it is less likely that they will buy the product. Past studies show that perceived risk is an important factor, which influences a consumer's purchase intention (Chang & Tseng, 2013; Grewal et al., 1994; Sweeney et al., 1999).

Mitchell (1999) contended that perceived risk is actually a "multidimensional phenomenon" which can be segmented into various risk components. A rich stream of consumer behavior literature supports the usage of these risk facets to understand consumer product and service evaluations and purchases. Reviews of perceived risk illustrate a variety of conceptualizations of perceived risk (Dowling, 1986; Dowling & Staelin, 1994). The more common components of perceived risk include performance, financial, social and psychological risks (Beneke et al., 2013; Chang & Tseng, 2013; Featherman & Pavlou, 2003; Jacoby & Kaplan, 1972). Financial risk is the potential monetary loss and is defined as the potential monetary outlay associated with the initial purchase price as well as the subsequent maintenance cost of the product (Chang & Tseng, 2013; Grewal et al., 1994; Lim, 2003). Performance risk means the possibility that a purchased product fails to provide the desired benefits or does not function properly (Bauer, 1960; Crespo, del Bosque, & de los Salmones Sánchez, 2009; Grewal et al., 1994; Lim, 2003). Performance risk is considered to be a quality risk (Kang & Kim, 2013) and is similar to the usefulness or functionality of products. Social risk is a consumer's concern about how other people perceive their shopping behavior and about the potential loss of status (Crespo et al., 2009; Kang & Kim, 2013; Lim, 2003). Consumers contemplate whether purchasing and wearing an apparel product may damage or reduce their image in the eyes of others (Kang & Kim, 2013). It is the possibility that consumers' shopping behavior is not accepted by other society members (Lim, 2003). Psychological risk refers to mental stress a consumer suffers due to shopping behavior, and it will have a negative effect on the consumer's peace of mind or self-perception (Crespo et al., 2009; Lim, 2003). Psychological risk could possibly damage one's self-image and cause potential loss of self-esteem (ego loss) from the frustration of not achieving a buying goal (Kang & Kim, 2013).

Shopping and purchasing is a way for young consumers to define themselves and to create their identity of their own decision making. Young consumers use apparel products to develop, maintain, or reinforce their self-concept and self-image (Belk, 1988; Chaplin & John, 2005; Sirgy, 1982). For apparel products, one important factor influencing college students' purchasing decision is whether the product is fashionable or in style. Hiller Connell's (2010) study showed that there was a general perception among consumers that environmentally preferable apparel is less stylish when compared with mainstream apparel. Chen and Burns (2006) indicated that consumers may not want to sacrifice fashion or colors (e.g., organic fibers are available only in earth tone, off-white colors) for the sake of the environment when making decisions about the purchases of textiles and clothing. Therefore, fashion risk is an important risk dimension for clothing products. In the present study, we included fashion risk as one component of consumer perceived risk toward sustainable clothing.

The intentions to switch to products made by more socially responsible manufacturers may be contingent on the product's quality, style, and price equaling that of other available products (Dickson, 2000). It is highly likely that when a person contemplates the purchase of sustainable clothing, the individual has feelings of "uncertainty," "discomfort," and/or "anxiety." This type of feeling is the result of the consumer's perception of risk. Findings from previous studies reveal that the perception of higher prices is a major barrier to purchasing environmentally friendly apparel (Gam, 2011; Hill & Lee, 2012; Hiller Connell, 2010). Chen and Burns (2006) reported that although consumers say that environmental aspects of textiles are important to them, some retailers found it is more difficult to sell organic and recycled textile products because of their higher prices over regular items. Hiller Connell (2010) also found that society's expectation for individual's appearance is another external barrier to eco-conscious apparel acquisition. Therefore, based on previous literature research, we generate the following hypotheses to examine the influence of perceived risks on consumer attitude and purchase intention.

H2: There is a negative influence of (a) financial risk, (b) performance risk, (c) psychological risk, (d) social risk, and (e) fashion risk on consumers' attitude toward sustainable clothing.
H3: There is a negative influence of (a) financial risk, (b) performance risk, (c) psychological risk, (d) social risk, and (e) fashion risk on consumers' willingness to buy sustainable clothing.

Schwartz Values

Schwartz defined a value as a belief pertaining to desirable end states or modes of conduct, that transcends specific situations, guides selection or evaluation of behavior, people, and events, and is ordered by importance relative to other values to form a system of value priorities (Schwartz, 1992, 1994; Schwartz & Bilsky, 1987). Based on universal requirements of human existence, Schwartz and Bilsky (1987) identified 56 values, and specified a set of dynamic relations among the motivational types of values in an integrated manner. Forty-five of the value items have demonstrated nearly equivalent meaning across over 60 countries around the world (Schwartz, 1992, 1994; Schwartz & Sagiv, 1995). The Schwartz's (1992) value theory defines ten broad value types according to the motivation that underlies each of them. These ten value types include power, achievement, hedonism, stimulation, self-direction, universalism, benevolence, tradition, conformity, and security (see Table 1). Studies with samples from many countries have established that these indexes have adequate internal reliability, temporal stability, and external validity and that scores are not contaminated by social desirability (Bardi & Schwartz, 2003). The pursuit of each value has psychological, practical, and social consequences that may conflict or may be congruent with the pursuit of other values (Bardi & Schwartz, 2003).

Values are abstract principles that are central to an individual's self-concept (Dickson, 2000; Smith, 1982). They act as guides for assessing situations and determining an individual's social and ideological positions (Rokeach, 1979; Smith,

Table 1 Definitions of types of values and the items that represent and measure them

Values	Definitions	Items
Power	Social status and prestige, control or dominance over people and resources	Social power, authority, wealth
Achievement	Personal success through demonstrating competence according to social standards	Successful, capable, ambitious, influential
Hedonism	Pleasure and sensuous gratification for oneself	Pleasure, enjoying life
Stimulation	Excitement, novelty, and challenge in life	Daring, a varied life, an exciting life
Self-direction	Independent thought and action-choosing, creating, exploring	Creativity, freedom, independent, curious, choosing own goals
Universalism	Understanding, appreciation, tolerance, and protection of the welfare of all people and of nature	Broad-minded, wisdom, social justice, equality, a world at peace, a world of beauty, unity with nature, protecting the environment
Benevolence	Preservation and enhancement of the welfare of people with whom one is in frequent personal contact	Helpful, honest, forgiving, loyal, responsible
Tradition	Respect, commitment, and acceptance of the customs and ideas that traditional culture or religion provides the self	Humble, accepting my portion in life, devout, respect for tradition, moderate
Conformity	Restraint of actions, inclinations, and impulses likely to upset or harm others and violate social expectations or norms	Politeness, obedient, self-discipline, honoring parents and elders
Security	Safety, harmony, and stability of society, of relationships, and of self	Family security, national security, social order, clean, reciprocation of favors

Source Bardi and Schwartz (2003)

1982). Values are important for understanding various social psychological phenomena and are believed to play the role of fundamental beliefs that direct or motivate our behaviors and decision making (Bardi & Schwartz, 2003). Bardi and Schwartz (2003) maintained that people may act in accordance with their values even when they do not consciously think about them; thus, values may operate outside of awareness but they are available for retrieval from memory. Values are relatively stable motivational characteristics of persons that change little during adulthood (Bardi & Schwartz, 2003). Thus, the following hypothesis was posited:

H4: There is a relationship between Schwartz value types and (a) consumer attitudes and (b) consumers' willingness to buy with regard to sustainable clothing.

Research Method

A structured questionnaire was designed based on a careful review of pertinent literature. Sustainable clothing, which is also called environmentally and socially responsible apparel products (Goworek et al., 2012; Harris, Roby, & Dibb, 2015), was defined at the beginning of the questionnaire, and examples of sustainable clothing products were provided. The survey was first developed in English. A translation and a back translation of the questionnaire were performed by two researchers who are fluent in both Chinese and English to ensure translation equivalence.

In terms of the measures of the variables, specifically, six items measuring consumers' sustainability knowledge about apparel products were obtained from Park and Kim (2016). Measures of perceived risks were compiled from the literature (Dickson, 2000; Kang & Kim, 2013). These items were measured using a five-point Likert-type scale (1 = "strongly disagree" to 5 = "strongly agree"). We measured values with the Schwartz Value Survey (1992). The survey listed 56 value items, each followed by a short definition in parentheses. Participants rated each value as a guiding principle in their own life on a five-point scale from 1 (not at all important) to 5 (extremely important). The forty-five value items in the survey that have demonstrated nearly equivalent meaning across over 60 nations around the world were used to index the ten values. Indexes are computed by averaging the importance ratings of the value items that represent each value, listed in Table 1. Four items of measuring consumer attitude were adapted from Chan (2001) on a five-point semantic differential scale. Three items adopted from Sweeney et al. (1999) were used to measure consumers' willingness to purchase sustainable clothing products on a five-point scale after modification to the research specific context. A sample of current college students from a large Chinese university in Shanghai, China, was recruited during March 2017.

Data Analysis and Results

Sample Characteristics

A total of 430 responses were received, and a total of 381 valid responses were used for this study. Table 2 includes the sample characteristics information. Around 83% of the responses were from sophomore or juniors. The participants ranged in age from 18 to 26 years (45.4% were 18–20, 45.9% were 21–23, and 6.3% were 24–26). About 75% were female, and about 94% of the valid responses were from students majored in the School of Business. In terms of the discretionary spending per month, 11.0% of the respondents spent less than 500 CNY (approximately US $73.5), 52.0% spent 500–1,499 CNY (approximately US$73.5–220.4), and 28.6%

Table 2 Sample characteristics

Demographic information	Sample	
	n	%
Gender		
Male	95	24.9
Female	286	75.1
College		
School of business	357	93.7
Others	24	6.2
Age		
18–20	173	45.4
21–23	175	45.9
24–26	24	6.3
27–30	5	1.3
Over 30	4	1.0
Academic status		
Freshman	10	2.6
Sophomore	180	47.2
Junior	135	35.4
Senior	15	3.9
Graduate student	40	10.5
Others	1	0.3
Discretionary spending per month		
Less than 500 CNY (less than US$73.5)	42	11.0
500–999 CNY (US$73.5–146.9)	97	25.5
1000–1499 CNY (US$147–220.4)	101	26.5
1500–1999 CNY (US$220.6–294)	60	15.7
2000–3000 CNY (US$294.1–441.2)	49	12.9
Over 3000 CNY (over US$441.2)	32	8.1
Missing	1	0.3

Note CNY—Chinese Yuan Renminbi; at the time of the survey, US$1 = 6.8 CNY

1,500–3,000 CNY (approximately US$220.6–441.2). Around 8.1% had over 3,000 CNY (over US$441.2) discretionary spending per month.

Preliminary Data Analysis

Preliminary data analysis was conducted using SPSS 23. Table 3 shows descriptive analysis results of Chinese college students' apparel sustainability knowledge. Young Chinese consumers have medium-level apparel sustainability knowledge (M = 2.45–2.95). The respondents were more informed about environmental issues

Table 3 Descriptive results of consumers' apparel sustainability knowledge

Apparel sustainability knowledge	Mean	SD
I am informed about environmental issues (e.g., eco-fashion, environmental impact of clothing manufacturing) in the fashion apparel manufacturing business	2.953^{A}	0.948
I am informed about child labor/sweatshop issues in the fashion apparel manufacturing business	2.871^{AB}	0.955
I am knowledgeable about social equity issues (e.g., working conditions of factory workers, fair wage for factory workers) in the fashion apparel business	2.751^{BC}	0.864
I understand the environmental impact of apparel products across the supply chain	2.625^{CD}	0.885
I am knowledgeable about apparel brands that sell environmentally-friendly or socially responsible apparel products	2.562^{D}	0.920
I know more about environmentally or socially-responsible apparel business than the average person	2.454^{D}	0.847

Note ABCD denotes group differences by post hoc analysis (Tukey HSD)

(e.g., eco-fashion, environmental impact of clothing manufacturing) and child labor/sweatshop issues in the fashion apparel manufacturing business than the sustainable apparel brands/products. The respondents did not perceive themselves as more knowledgeable about socially responsible apparel business than the average person.

Table 4 shows the descriptive analysis results of the ten value types. Hedonism, security, and self-direction are the three values rated higher than other value types. Tradition, stimulation, and power were rated lower among the ten value types. Young Chinese consumers are willing to spend time and money in pleasure and sensuous gratification for themselves; they put high priority for safety, harmony, and stability, and they value independent thought and creativity, and freedom.

Table 4 Descriptive analysis results of ten value types

Value types	Mean	SD
Hedonism	4.243^{A}	0.639
Security	4.238^{A}	0.497
Self direction	4.100^{AB}	0.547
Conformity	4.055^{B}	0.614
Universalism	4.020^{B}	0.557
Benevolence	4.004^{B}	0.622
Achievement	3.754^{C}	0.658
Tradition	3.698^{C}	0.576
Stimulation	3.537^{D}	0.757
Power	3.290^{E}	0.724

Note ABCDE denotes group differences by post hoc analysis (Tukey HSD)

Due to the multidimensional nature of perceived risk, exploratory factor analysis (EFA) was first conducted for perceived risk to investigate the dimensional structure of the variable. The KMO measure of 0.816 and the Bartlett's test statistic ($p < 0.000$) indicate that the data is appropriate for factor analysis. The EFA solution was determined using the following criteria: eigenvalue (>1), variance explained by each component, scree plot, loading score for each factor (⩾0.5), and meaningfulness of each dimension. Items were eliminated if they showed low communalities (<0.30), low factor loadings (<0.40), and/or high cross-loadings (>0.40). As a result, one item of the financial risk factor and one item of the performance risk factor were eliminated. After the eliminations, the final EFA resulted in a five-factor model with 14 items, including 3 items of social risk, 3 items of fashion risk, 3 items of performance risk, 3 items of psychological risk, and 2 items of financial risk. Table 5 includes the EFA results. The result is consistent with previous studies on multidimensional nature of perceived risk. Since the internal reliability Cronbach's alpha values for social risk (alpha = 0.890), fashion risk (alpha = 0.858), performance risk (alpha = 0.853), and psychological risk (alpha = 0.842) are satisfactory, the study calculated the averages from the respective items for social risk, fashion risk, performance risk, and psychological risk in the further analysis. The internal reliability Cronbach's alpha for financial risk is low (alpha = 0.450) although the two items form one factor. This result indicates that the present study does not confirm Kang and Kim (2013)'s financial risk scale. In the further regression analysis, we included all the three items related to financial risk, reflecting consumer's perception of purchasing cost, maintenance cost, and risk of wasting money.

EFA was also conducted on apparel sustainability knowledge, consumer attitudes, and willingness to buy. The results revealed unidimensionality in sustainability knowledge (internal reliability Cronbach's alpha = 0.832), consumer attitude (internal reliability Cronbach's alpha = 0.855), and willingness to buy (internal reliability Cronbach's alpha = 0.825).

Effects of Apparel Sustainability Knowledge and Perceived Risk on Consumer Attitude and Willingness to Buy

To examine the effects of apparel sustainability knowledge and perceived risk on consumer attitude and willingness to buy, standard multiple regression analyses were performed with consumer attitude and willingness to buy as the dependent variables and consumer apparel sustainability knowledge and perceived risks as the independent variables. We created composite variables by averaging the respective items for apparel sustainability knowledge, social risk, fashion risk, performance risk, psychological risk, and included those composite variables and the three items related to the financial risk (purchasing cost, maintenance cost, and risk of wasting money) as the independent variables in the regression analysis. The average of the

Table 5 EFA results of perceived risks

	Social risk	Fashion risk	Performance risk	Psychological risk	Financial risk
I am worried about what others will think of me when I purchase the ESRAP	**0.863**	0.176	0.151	0.159	0.096
I am worried that my friends might think I look weird or funny in the ESRAP	**0.830**	0.251	0.182	0.187	0.037
I will not feel comfortable wearing the ESRAP in public	**0.802**	0.289	0.163	0.216	0.062
I will feel the ESRAP I purchased might be outdated	0.162	**0.901**	0.063	0.155	0.029
I will feel the ESRAP I purchased might not be the latest style	0.219	**0.802**	0.083	0.239	0.068
I will feel that the ESRAP I purchased might not be in fashion	0.308	**0.767**	0.121	0.156	0.077
The quality of the ESRAP will be unsatisfactory	0.142	0.094	**0.871**	0.147	0.102
The quality of the ESRAP will be poor	0.124	0.046	**0.860**	0.134	0.040
The life cycle of the ESRAP will not be long	0.149	0.098	**0.837**	0.029	0.030
The ESRAP that I purchased will not look good on me	0.128	0.141	0.123	**0.872**	0.001
It will be difficult for me to be able to match the ESRAP with my current clothing	0.130	0.246	0.096	**0.852**	0.002
Purchasing ESRAP will not match my own personal image	0.366	0.166	0.126	**0.708**	0.155
It will cost so much for me to purchase the ESRAP	0.115	0.174	−0.015	−0.064	**0.797**
It will cost a lot to manage and keep the ESRAP in good shape	0.020	−0.051	0.149	0.148	0.789
Eigenvalues	2.495	2.380	2.367	2.257	1.318
% of variance	17.821	17.003	16.905	16.118	9.412
Cumulative variance %	17.821	34.823	51.728	67.847	77.259
Cronbach's alpha	0.890	0.858	0.853	0.842	0.450

Note Rotated factor loadings that are 0.4 or larger are set in bold; ESRAP stands for the environmentally and socially responsible apparel products

items of consumer attitudes and the average of the items measuring willingness to buy were individually used as the dependent variable. Durbin-Watson statistic is in the range of 1.5–2.5, and the variance inflation factor (VIF) is under 4.0, indicating the data is appropriate for regression analysis.

As shown in Table 6, regression analysis results revealed that consumer attitude and willingness to buy were significantly predicted by the regression models. Consumer apparel sustainability knowledge positively and significantly predicts both consumer attitude and willingness to buy. Therefore, H1 (a) and H1 (b) were supported. Performance risk, risk of wasting money, and fashion risk negatively and significantly predict consumer attitude, which means when consumer perceives higher level of performance risk, risk of wasting money, and fashion risk with regard to purchasing sustainable clothing, they will have more negative attitude toward sustainable clothing. Other aspects of perceived risk such as psychological risk, social risk, and purchasing cost do not have a strong relationship with consumer attitude. Performance risk, risk of wasting money, and social risk negatively and significantly predict consumer willingness to buy, which shows when consumer perceives higher level of performance risk, social risk, and risk of wasting money, they are less willing to buy sustainable clothing. Other aspects of perceived risk such as psychological risk, fashion risk, and purchasing cost do not have a strong relationship with consumer purchase intention. Thus, H2 and H3 were partially supported.

Effect of Values on Consumer Attitude and Willingness to Buy

The ten value types from Schwartz's value model were used for the regression analyses. We created composite variables by averaging the respective items for each variable: ten value types, consumer attitudes, and willingness to buy. Standard multiple regression analyses were performed with consumer attitude and willingness to buy individually as the dependent variable and the ten value types as the independent variables (power, achievement, hedonism, stimulation, self-direction, universalism, benevolence, tradition, conformity, and security). Regression analysis results show the Durbin-Watson statistic is in the range of 1.5–2.5 and the variance inflation factor (VIF) is under 4.0, indicating the data is appropriate for regression analysis. As shown in Table 7, regression analysis results revealed that consumer attitude and willingness to buy are significantly predicted by their personal value. Specifically, the value type benevolence significantly and positively predicts consumer attitude. Benevolence strongly and positively predicts consumer willingness to buy, while the value type achievement significantly and negatively predicts consumer willingness to buy. Thus, H4 was partially supported.

Table 6 Results of multiple regression

Dependent variables	Independent variables	Durbin-Watson statistic	Adj. R^2	F	Standardized beta	t-value
Consumer attitude		1.822	0.131	8.134		
	Sustainability knowledge				0.117	2.388**
	Performance risk				−0.179	−3.226***
	Psychological risk				−0.068	−1.172
	Social risk				−0.037	−0.586
	Fashion risk				−0.120	−1.994**
	Purchasing cost				0.060	1.132
	The risk of wasting money				−0.136	−2.467**
	Maintenance cost				0.122	2.374**
Willingness to buy		1.797	0.204	13.206		
	Sustainability knowledge				0.188	3.994****
	Performance risk				−0.106	−1.994**
	Psychological risk				−0.062	−1.119
	Social risk				−0.181	−2.995***
	Fashion risk				−0.070	−1.219
	Purchasing cost				0.054	1.064
	The risk of wasting money				−0.209	−3.967****
	Maintenance cost				0.085	1.728*

*$P < 0.10$; **$P < 0.05$; ***$P < 0.01$; ****$P < 0.001$

Table 7 Results of multiple regression

Dependent variables	Independent variables	Durbin-Watson statistic	Adj. R^2	F	Standardized beta	t-value
Consumer attitude		1.835	0.072	3.913		
	Power				−0.041	−0.649
	Achievement				0.020	0.252
	Hedonism				−0.057	−1.002
	Stimulation				−0.000	−0.001
	Self-direction				0.066	0.827
	Universalism				0.041	0.453
	Benevolence				0.197	2.261**
	Tradition				−0.023	−0.293
	Conformity				0.011	0.139
	Security				0.081	1.076
Willingness to buy		1.781	0.101	5.245		
	Power				0.103	1.677*
	Achievement				−0.210	−2.679***
	Hedonism				−0.048	−0.851
	Stimulation				−0.020	−0.309
	Self-direction				0.111	1.410
	Universalism				0.158	1.794*
	Benevolence				0.296	3.451***
	Tradition				−0.094	−1.211
	Conformity				−0.052	−0.646
	Security				0.062	0.840

*$P < 0.10$, **$P < 0.05$, ***$P < 0.01$

Discussion and Implications

A limited number of academic studies have examined a holistic view of young consumers' knowledge level and attitude toward purchasing apparel products with different environmental and social responsibility attributes—organic, fair trade, and recycled (Hwang et al., 2015). Although more and more apparel companies have been concerned with the integration of environmental, social, and ethical considerations into business strategies and practices, the Chinese consumer's sustainability knowledge about apparel industry and Chinese consumer's perceptions of sustainable clothing have not been sufficiently examined in academia. Thus, this study investigated Chinese post-90s' perceptions of sustainable clothing by surveying Chinese college students. The study gained perspectives from Chinese post-90s' college students regarding their apparel sustainability knowledge, perceived risk of sustainable clothing, personal values. Moreover, it explored how the young Chinese consumers' apparel sustainability knowledge, personal values, and perceived risks impact their attitude toward and willingness to buy sustainable clothing products. This study provides valuable baseline information for the textile and apparel industry.

Previous studies suggested that consumers expressed positive sentiments toward sustainability in general but they lacked knowledge on social responsible practices in the apparel industry (Gam & Banning, 2011; Hill & Lee, 2012; Hiller Connell, 2010; Hwang et al., 2015). Dickson (2000) suggested that greater knowledge about industry issues leads to greater concern for industry workers, and when consumers have greater knowledge and concern, their support for socially responsible business goals increases. In the present study, Chinese young consumers may have insufficient knowledge of sustainability issues in the apparel industry and they slightly disagree that they have much knowledge. This confirms previous research stating that environmentally and socially responsible consumer behavior may not be catching on because educating consumers has not been a priority of industry (Dickson, 2000). Since the environmental and social issues in the apparel industry have only recently reached prominence in the media in China, the typical clothing consumer may feel inadequately informed or confused about actual industry conditions. Among all the sustainability issues in the apparel industry, young consumers are more aware of environmental impact and child labor issues than sustainable clothing brands or business, indicating the need for strengthening and expanding marketing promotion for sustainable apparel brands and business in China. The study results show apparel sustainability knowledge strongly and positively predicts consumer attitude and willingness to buy, confirming previous research on the importance of sustainability knowledge (Dickson, 2000; Hill & Lee, 2012; Hwang et al., 2015). Thus, knowledge underlies attitudes consumers form about sustainable apparel business. If greater levels of knowledge can be achieved, support for socially and environmentally responsible businesses will increase to a point that more directly affects purchase behavior. Firms in the textile and apparel industry may need to focus on providing information to consumers in hopes of strengthening their support for eco-friendly and socially responsible business. Chinese apparel firms and retailers could develop some educational strategies in their marketing and branding effort for sustainable apparel products. For example, the literature provided ideas such as using hangtags on garments in retail stores and extensive editorial space in catalogs to inform consumers about desired product attributes as well as company engagement in prosocial initiatives (Dickson, 2000; Hyllegard et al., 2014).

Sustainable clothing is still a new concept for many Chinese young consumers, and sustainable clothing business is an emerging market in China. Identified as a critical determinant of consumers' willingness to buy a new item, such as sustainable clothing product, is the perceived risk associated with the purchase (Grewal et al., 1994). The study findings reveal that performance risk, maintenance risk, and risk of wasting money significantly influence consumer attitude, and fashion risk impacts attitude to some extent. Performance risk, social risk, and risk of wasting money significantly impact consumer willingness to buy sustainable clothing, and maintenance cost impacts willingness to buy to some extent. Psychological risk and purchasing cost are not influential predictors for consumer attitude and willingness to buy. The results indicate that consumers are not concerned about the purchasing cost or psychological uncertainty of sustainable clothing; however, the consumers are more concerned about whether a purchased sustainable apparel product fails to provide the desired benefits or does not function properly, and whether there is a potential cost for maintaining sustainable clothing, and whether there is feeling of wasting money for purchasing sustainable clothing. On the other hand, consumers may not feel restricted by the products offered by environmentally and socially responsible businesses; they may simply be overwhelmed by the complexity of product attributes that can be used in decisions to purchase clothing. Considering the fact that sustainable clothing products are relatively new product category, consumer may not fully understand the attributes of sustainable clothing; they are not sure about sustainable clothing's quality and performance and may merely think purchasing sustainable clothing needs extra maintenance cost or there is high possibility of wasting money.

Educating consumers seems to be an essential component for directing their purchases toward environmentally and socially responsible businesses. However, education alone will unlikely command changes in consumer behavior (Dickson, 2000). Manufacturers and retailers will have to supply the products consumers want, along with the assurance the business behaves responsibly. Purchasing sustainable clothing involves a variety of possible risks, including performance risk, psychological risk, social risk, fashion risk, higher purchase cost, maintenance cost, and risk of wasting money. Consumer's perception of the uncertainty and adverse consequences of engaging in sustainable purchasing influences their purchase intention. Moreover, it is necessary to know what types of perceived risk impact consumer attitude and purchase intention most.

In terms of the relationship between values and consumer attitude and willingness to buy sustainable clothing products, benevolence value significantly and positively impacts consumer attitude and purchase intention; achievement value significantly and negatively impacts consumer willingness to buy. Universalism impacts consumer willingness to buy to some extent. These results indicate that consumers who value benevolence more will have more favorable attitude toward sustainable clothing and more willing to buy sustainable clothing.

Limitations and Future Research Directions

There are several limitations in the study which provide opportunities for future research. First, generalization of the research findings is limited because of the use of a convenience sample (current college students) within a limited geographical location. Caution should be used in generalizing the findings of this study to the Chinese post-90s' population as a whole. The sample is also biased in that it reflects a more highly educated portion of the Chinese post-90s. Less-educated Chinese post-90s' thoughts on environmental and social responsibility may not be represented here. Future research may use a random sample that is more heterogeneous in terms of geographic location and educational level to confirm the findings. Second, this study examined Chinese young consumers' sustainable clothing consumer behaviors. Future study could be extended to other countries to investigate how young consumers in other countries perceive sustainable clothing. Cross-cultural research is desirable for research in sustainable consumer behavior. Third, this study surveyed post-90s; however, future research should be conducted with other generational cohorts too (e.g., post-70s and post-80s) to understand the way different generational cohorts perceive sustainable apparel products and to further investigate the systematic relationships among beliefs, attitudes, and purchase intentions. Fourth, financial risk in this study needs future scale development and the three items used in the study do not represent the financial risk well. Thus, more future research is needed for the financial risk scale. Finally, in order to obtain more in-depth reasoning behind Chinese post-90s' consumer behavior for the sustainable products, focus group interviews would be alternative approaches to explore in future research.

References

Ajzen, I., & Fishbein, M. (1980). *Understanding attitudes and predicting social behavior*. Englewood Cliffs, NJ: Prentice-Hall.

Bardi, A., & Schwartz, S. H. (2003). Values and behavior: Strength and structure of relations. *Personality and Social Psychology Bulletin, 29*(10), 1207–1220.

Bauer, R. A. (1960). Consumer behavior as risk taking. In R. S. Hancock (Ed.), *Dynamic marketing for a changing world* (pp. 389–398). Chicago: American Marketing Association.

Belk, R. W. (1988). Possessions and the extended self. *Journal of Consumer Research, 15*(2), 139–168.

Beneke, J., Flynn, R., Greig, T., & Mukaiwa, M. (2013). The influence of perceived product quality, relative price and risk on customer value and willingness to buy: A study of private label merchandise. *Journal of Product & Brand Management, 22*(3), 218–228.

Bhaduri, G., & Ha-Brookshire, J. E. (2011). Do transparent business practices pay? Exploration of transparency and consumer purchase intention. *Clothing and Textiles Research Journal, 29*(2), 135–149.

Butler, S. M., & Francis, S. (1997). The effects of environmental attitudes on apparel purchasing behavior. *Clothing and Textiles Research Journal, 15*(2), 76–85.

Cerini, M. (March 14, 2016). Shanghai startup sets an eco-friendly model for China's clothing wholesalers. Retrieved from https://www.forbes.com/sites/mariannacerini/2016/03/14/shanghai-start-up-sets-an-eco-friendly-model-for-chinas-clothing-wholesalers/#2aaffdc71db8.

Chan, R. Y. K. (2001). Determinants of Chinese consumers' green purchase behavior. *Psychology and Marketing, 18*(4), 389–413.

Chang, E.-C., & Tseng, Y.-F. (2013). Research note: E-store image, perceived value and perceived risk. *Journal of Business Research, 66*(7), 864–870.

Chaplin, L. N., & John, D. R. (2005). The development of self-brand connections in children and adolescents. *Journal of Consumer Research, 32*(1), 119–129.

Chen, H.-L., & Burns, L. D. (2006). Environmental analysis of textile products. *Clothing and Textiles Research Journal, 24*(3), 248–261.

Crespo, Á. H., del Bosque, I. R., & de los Salmones Sánchez, M. M. G. (2009). The influence of perceived risk on Internet shopping behavior: A multidimensional perspective. *Journal of Risk Research, 12*(2), 259–277.

Dickson, M. A. (2000). Personal values, beliefs, knowledge, and attitudes relating to intentions to purchase apparel from socially responsible businesses. *Clothing and Textiles Research Journal, 18*(1), 19–30.

Dickson, M. A., Loker, S., & Eckman, M. (2009). *Social responsibility in the global apparel industry*. New York: Fairchild Books.

Dowling, G. R. (1986). Perceived risk: The concept and its measurement. *Psychology and Marketing, 3*(3), 193–210.

Dowling, G. R., & Staelin, R. (1994). A model of perceived risk and intended risk handling activity. *Journal of Consumer Research, 21*(1), 119–134.

Featherman, M. S., & Pavlou, P. A. (2003). Predicting e-services adoption: A perceived risk facets perspective. *International Journal of Human-Computer Studies, 59*(4), 451–474.

Gam, H. J. (2011). Are fashion-conscious consumers more likely to adopt eco-friendly clothing? *Journal of Fashion Marketing and Management: An International Journal, 15*(2), 178–193.

Gam, H. J., & Banning, J. (2011). Addressing sustainable apparel design challenges with problem-based learning. *Clothing and Textiles Research Journal, 29*(3), 202–215.

Gam, H. J., Cao, H., Farr, C., & Kang, M. (2010). Quest for the eco-apparel market: A study of mothers' willingness to purchase organic cotton clothing for their children. *International Journal of Consumer Studies, 34*(6), 648–656.

Goworek, H., Fisher, T., Cooper, T., Woodward, S., & Hiller, A. (2012). The sustainable clothing market: An evaluation of potential strategies for UK retailers. *International Journal of Retail & Distribution Management, 40*(12), 935–955.

Grewal, D., Gotlieb, J., & Marmorstein, H. (1994). The moderating effects of message framing and source credibility on the price-perceived risk relationship. *Journal of Consumer Research, 21*(1), 145–153.

Harris, F., Roby, H., & Dibb, S. (2015). Sustainable clothing: challenges, barriers and interventions for encouraging more sustainable consumer behaviour. *International Journal of Consumer Studies, 40,* 309–318.

Hill, J., & Lee, H. H. (2012). Young Generation Y consumers' perceptions of sustainability in the apparel industry. *Journal of Fashion Marketing and Management: An International Journal, 16*(4), 477–491

Hiller Connell, K. Y. (2010). Internal and external barriers to eco-conscious apparel acquisition. *International Journal of Consumer Studies, 34*(3), 279–286.

Hiller Connell, K. Y. (2011). Exploring consumers' perceptions of eco-conscious apparel acquisition behaviors. *Social Responsibility Journal, 7*(1), 61–73.

Hwang, C. G., Lee, Y.-A., & Diddi, S. (2015). Generation Y's moral obligation and purchase intentions for organic, fair-trade, and recycled apparel products. *International Journal of Fashion Design, Technology and Education, 8*(2), 97–107.

Hyllegard, K. H., Ogle, J. P., & Yan, R.-N. (2014). College students' responses to prosocial marketing claims on apparel hang tags. *Journal of Fashion Marketing and Management: An International Journal, 18*(3), 269–283.

Jacoby, J., & Kaplan, L. B. (1972). *The components of perceived risk.* Paper Presented at the Proceedings of the Third Annual Conference of the Association for Consumer Research.

Joergens, C. (2006). Ethical fashion: Myth or future trend? *Journal of Fashion Marketing and Management, 10*(3), 360–371.

Kang, J., & Kim, S.-H. (2013). What are consumers afraid of? Understanding perceived risk toward the consumption of environmentally sustainable apparel. *Family and Consumer Sciences Research Journal, 41*(3), 267–283.

Kolk, A., Hong, P., & Van Dolen, W. (2010). Corporate social responsibility in China: An analysis of domestic and foreign retailers' sustainability dimensions. *Business Strategy and the Environment, 19*(5), 289–303.

Lim, N. (2003). Consumers' perceived risk: Sources versus consequences. *Electronic Commerce Research and Applications, 2*(3), 216–228.

Ma, Y. J., & Lee, H.-H. (2012). Understanding consumption behaviours for fair trade non-food products: Focusing on self-transcendence and openness to change values. *International Journal of Consumer Studies, 36*(6), 622–634.

McDonald, S., Oates, C., Thyne, M., Alevizou, P., & McMorland, L.-A. (2009). Comparing sustainable consumption patterns across product sectors. *International Journal of Consumer Studies, 33*(2), 137–145.

Mitchell, V. W. (1999). Consumer perceived risk: Conceptualisations and models. *European Journal of Marketing, 33*(1/2), 163–195.

O'Cass, A., & Choy, E. (2008). Studying Chinese generation Y consumers' involvement in fashion clothing and perceived brand status. *Journal of Product & Brand Management, 17*(5), 341–352.

O'Cass, A., & Siahtiri, V. (2014). Are young adult Chinese status and fashion clothing brand conscious? *Journal of Fashion Marketing and Management: An International Journal, 18*(3), 284–300.

Park, H., & Kim, Y.-K. (2016). Proactive versus reactive apparel brands in sustainability: Influences on brand loyalty. *Journal of Retailing and Consumer Services, 29*, 114–122.

Peter, J. P., & Ryan, M. J. (1976). An investigation of perceived risk at the brand level. *Journal of Marketing Research, 13*(2), 184–188.

Rapp, J. (2016). Eco-fashion gets hip for China's 'light green' consumers. Retrieved from https://jingdaily.com/eco-fashion-gets-hip-for-chinas-light-green-consumers/.

Rokeach, M. (1979). *Understanding human values: Individual and societal*. New York: Free Press.

Schwartz, S. H. (1992). Universals in the content and structure of values: Theoretical advances and empirical tests in 20 countries. In P. Z. Mark (Ed.), *Advances in experimental social psychology* (Vol. 25, pp. 1–65). New York: Academic Press.

Schwartz, S. H. (1994). Are there universal aspects in the structure and contents of human values? *Journal of Social Issues, 50*(4), 19–45.

Schwartz, S. H., & Bilsky, W. (1987). Toward a universal psychological structure of human values. *Journal of Personality and Social Psychology, 53*(3), 550.

Schwartz, S. H., & Sagiv, L. (1995). Identifying culture-specifics in the content and structure of values. *Journal of Cross-cultural Psychology, 26*(1), 92–116.

Simões, F. (2016). Consumer behavior and sustainable development in China: The role of behavioral sciences in environmental policymaking. *Sustainability, 8*(9), 897.

Sirgy, M. J. (1982). Self-concept in consumer behavior: A critical review. *Journal of Consumer Research, 9*(3), 287–300.

Smith, M. J. (1982). *Persuasion and human action: A review and critique of social influence theories*. Belmont, CA: Wadsworth Publishing Company.

Smith, N. C. (2003). Corporate social responsibility: Whether or how? *California Management Review, 45*(4), 52–76.

Song, M. (April 24, 2017). In China, making a fashion statement. Retrieved from http://blogs.ei.columbia.edu/2017/04/24/in-china-making-a-fashion-statement/.

Sweeney, J. C., Soutar, G. N., & Johnson, L. W. (1999). The role of perceived risk in the quality-value relationship: A study in a retail environment. *Journal of Retailing, 75*(1), 77–105.

U.S. Census Bureau. (2017). International Data Base. Retrieved July 3, 2017 https://www.census.gov/population/international/data/idb/informationGateway.php.

Application of Motivation-Opportunity-Ability Theory in the Consumption of Eco-fashion Products: Were Chinese Consumers Underestimated?

Ruirui Zhang and Chunmin Lang

Abstract Fueled by increased affluence and modernization process, China is experiencing dramatic changes in its culture, social value, and consumption values. For Chinese people, the satisfaction of functional needs has been taken place by social, emotional, and experiential needs in apparel consumption. This transformation indicates that Chinese consumers' attitude and products selection are moving toward health, happiness, and self-satisfaction. In general, Chinese have a stronger financial incentive to live sustainably and they have more flexibility to change. Therefore, the purpose of the study is to determine the factors that influence Chinese consumers' intention to pursue eco-fashion consumption. The Motivation-Opportunity-Ability theory was applied as the theoretical foundation in this study. The data were collected in China through an online survey. Participants (n = 410) were recruited through email survey link and the social media app (WeChat). The SEM analysis results indicated that self-identity, environment consciousness, environmental apparel knowledge, and attitude significantly determine Chinese consumers' intention to pursue eco-fashion consumption. In addition, the perceived opportunity was found to have a direct effect on the intention toward eco-fashion consumption, as well as mediate the relationship between perceived ability and intention toward eco-fashion consumption.

Keywords Eco-fashion consumption · Motivation-Opportunity-Ability theory
Self-identity · Environmental consciousness · Environmental apparel knowledge

---

R. Zhang (✉)
Department of Fashion Design and Retailing, Framingham State University,
234A Hemenway Annex, Framingham, MA 01701, USA
e-mail: rzhang@framingham.edu

C. Lang
Department of Textiles, Apparel Design, & Merchandising, College of Agriculture,
Louisiana State University, 143 Human Ecology, Baton Rouge, LA 70803, USA
e-mail: cmlang@lsu.edu

Y. Xu et al. (eds.), *Chinese Consumers and the Fashion Market*, Springer Series in Fashion Business, https://doi.org/10.1007/978-981-10-8429-4_6

Introduction

Fueled by increased affluence and modernization processes, China is experiencing dramatic changes in its culture, social values, and consumption practices. Xiao and Kim (2009) revealed that, for Chinese people, the satisfaction of functional needs has been replaced by social, emotional, and experiential needs in apparel consumption. This transformation indicates that Chinese consumers' attitude and product selection are moving toward health, happiness, and self-satisfaction. In addition, due to higher educational levels, more disposable income, increasing social responsibility and environmental awareness, Chinese consumers are evolving in their behaviors and practices and are beginning to promote sustainable consumption. Brodwin (2015) found that Chinese consumers have shown more willingness to buy environmentally friendly products when compared to US consumers. In general, the Chinese have a stronger financial incentive to live sustainably and they have more flexibility to change.

Although a variety of studies have been done on sustainable consumption in developed countries, a deeper understanding of Chinese consumers' sustainable consumption is lacking. Environmental consciousness was not a central theme in China since the Chinese government's highest priority unquestionably was economic development in the past thirty years. Even though the widespread environmental problems, such as water pollution, air pollution, and toxic waste releasing, have been inevitably discerned, economic growth and wealth creation are still the most valuable goals and are always given a higher level of importance (Harris, 2006). Due to a lack of knowledge in ecology and education regarding environmental issues during the rapid economic development period from 1979 to 2000, Chinese consumers' environmental consciousness has been insufficient; especially those Chinese who were living outside the major urban areas.

The aim of this study is to address existing research gaps in addition to delivering practical formation for sustainable development and consumption in the context of fashion business in China. The purpose of this study is also to (1) develop a conceptual model based on Motivation-Opportunity-Ability (MOA) theory to examine Chinese eco-fashion consumption behaviors; (2) identify antecedents, such as Chinese consumers' man–nature-oriented environmental consciousness and attitude, the level of environmental knowledge, and green self-identity effects on eco-fashion consumption; (3) to determine the two external factors, opportunities and ability, and their effects on eco-fashion consumption.

Literature Review and Hypotheses Development

Eco-fashion Consumption

Eco-fashion, one of the green versions of consumer goods, is defined as the type of clothing that is designed and produced to maximize benefits for people and society and to minimize environmental impacts (Claudio, 2007; Joergens, 2006; Ochoa,

2011). The clothing is made of biodegradable and recycled materials and/or produced through environmentally friendly sourcing, production, and distribution and/or is involved with corporate socially responsible practices. This is what is considered eco-fashion (Chan & Wong, 2012; Claudio, 2007; Young, Hwang, McDonald, & Oates, 2010). Therefore, eco-fashion emphasizes the environmentally responsible aspect instead of the socially responsible aspect of a business.

Many green versions of consumer products are now widely available in China. Ye (2000) mentioned that over 75% of Chinese consumers have strongly expressed their intention toward buying green products. Chan (2000) discovered that Chinese consumers are very familiar with green consumer products such as eco-friendly detergents and batteries. He also found that one-third of participants in his study bought at least one of those products in the past six months. Shen, Wang, Lo and Shum (2012) found that Chinese consumers are willing to pay a premium price for clothing which is made in an environmentally and socially responsible way. Previous studies (Chan & Wong, 2012; Niinimäki, 2010) found that product-related attributes (e.g., product design, quality, and price) of eco-fashion and store-related attributes (e.g., store design, environment, ethical practices, and shop convenience) of eco-fashion are two important predictors of eco-fashion consumption behaviors. Chan and Wong (2012) specifically claimed that among Chinese consumers, store-related attributes determine Chinese consumers' attitude, which in turn influences their eco-fashion consumption behaviors. They also suggested that the price of eco-fashion, as a critical factor, had a major effect on Chinese consumers' eco-fashion consumption and moderated the relationship between store-related attributes and eco-fashion consumption. The price factor should not be underestimated in further Chinese consumer behavioral studies.

In addition to those external influences motivating environmental consciousness, the internal traditional Chinese cultural values may also help to explain why eco-fashion consumption behaviors are progressively becoming remarkable among Chinese consumers. Chan, Wong, and Leung (2008) bring a unique perspective to explain Chinese ethical behaviors. They used the terminology "man–nature orientation" that Kluckhohn and Strodtbeck created in 1961 in explaining Chinese ethical consumption behaviors. They pointed out that the "man–nature orientation" is similar to "Chinese place great emphasis on living in harmony with Nature" (Chan, 2000; Chan et al., 2008; Yau, 1988). This philosophy was funded by the thought of *Lao Tzu*. This great philosopher, who lived long before the time of Confucius, taught that in order to achieve a "man–nature unity," the human beings, as a part of nature, should not master but should adapt to or integrate with the natural world. This Chinese "Man–nature orientation" is akin to Western literature's eco-centric orientation (Dunlap et al., 2000; Dunlap & Jones, 2002). Therefore, if Chinese consumers in the current society adhere to this traditional cultural value, they are more likely to adopt environmentally responsible consumption behaviors due to this genetic and inherent righteousness (Chan et al., 2008).

Motivation-Opportunity-Ability (MOA) Theory

Motivation-Opportunity-Ability (MOA) theory was proposed by MacInnis and Jaworski in 1989 and it has been widely employed to explain consumer behaviors and knowledge management practices (MacInnis, Moorman, & Jaworski, 1991; Siemsenm, Roth, & Balasubramanian, 2008; Argote, McEvily, & Reagans, 2003). This theory suggests that motivation, opportunity, and ability determine the degree to which individuals process information. It also particularly highlights the importance of the motivation factor that acts as a driver of behaviors. In general, motivation is a force that directs individuals toward goals and determines the individuals' willingness and readiness to act. Marketing research usually investigates ways which can be used to motivate consumers' engagement in their decision-making journeys (Gruen, Osmonbekov, & Czaplewski, 2006). Opportunity shows the contextual or environmental mechanisms which are conducive to achieving desired goals or behavioral outcomes. There are several situational factors, such as time, opportunity, availability, attention paid, and a number of repetitions which will trigger or deter the desired goals and behavioral outcomes (MacInnis & Jaworski, 1989). In the MOA theory, ability represents a person's skill or proficiencies in certain behaviors. In marketing and consumer studies, the accessibility and availability of situational factors provide the foundation for processing ability. Taking the position that (1) Chinese consumers' eco-fashion consumption intention as a behavioral outcome, (2) environmental consciousness, self-identity, consumers' attitude, and environmental apparel knowledge are proposed to be motivational factors, and (3) perceived opportunity and ability are conducive to eco-fashion consumption intention, we propose that the MOA theory can serve as an explanatory theoretical framework in the context of this study.

Motivations in the MOA Theory

In behavioral science studies, motivation is usually considered as a force which leads people toward goals and behavioral action. In marketing studies, companies develop different psychological and physical cues through which consumers can be motivated to engage in processing information and/or buying intention and/or behavior (Gruen et al., 2006). Motivation has been defined as goal-directed arousal (Park & Mittal, 1985). In this study, the goal of arousal is classified as willingness to protect the environment for the future (environmental sustainability and consciousness); awareness of the harmful effects of apparel production processes (environmental apparel knowledge); having positive attitudes toward buying eco-fashion (attitude), and a desire to be an "environmental consumer" (self-identity). High motivation in those aspects implies that Chinese consumers are willing to buy eco-fashion. Previous studies found that green self-identification motivates consumers' eco-fashion consumption (Chan & Wong, 2012; Niinimäki,

2010). Consumers' high level of environmental apparel knowledge and green attitude predicts their eco-fashion consumption behaviors (Hiller Connell & Kozar, 2012; Johe & Bhullar, 2016). This study identifies four variables, including environmental consciousness, self-identity, environmental apparel knowledge, and attitudes, as motivational variables in the proposed eco-fashion consumption conceptual model.

Environmental Consciousness

Environmental consciousness is a multi-dimensional construct which comprises cognitive, attitudinal, and behavioral components (Schelgelmilch, Bohlen, & Diamontopoulos, 1996). It is defined as a psychological tendency of performing in pro-environmental behaviors. Individuals' value judgments, recognition, and behavioral intention toward environmental issues are often reflected in the level of one's environmental consciousness (Chang, 2012; Schlegelmilch et al., 1996; Zelezny & Schultz, 2000; Zheng, 2010). Pro-environmental behaviors, such as signing a petition to support stricter environmental laws, participating in a protest against a company that is harming the environment, writing a letter to public officials to increase their support of environmental protection efforts, etc., are highly correlated with environmental consciousness (Cordano, Welcomer, & Scherer, 2003). Eco-consumption behaviors, such as using a product that is durable and safe for the environment, or buying clothing with recycled content, are psychologically engaged with the level of one's environmental consciousness. Consumers who have a low level of environmental consciousness are less likely to buy environmental products because they do not consider environmental issues as critical factors in their decision-making process. On the contrary, those with a high awareness of environmental consciousness are more likely to adopt a wider range of environmentally conscious behaviors (Schlegelmilch et al., 1996).

Unfortunately, due to the holding of traditional views regarding Chinese' social, economic, and political structure, many people maintain that the Chinese have far less environmental consciousness than Western consumers. However, Chan (1999) found that not only have Chinese residents shown their actual commitment toward eco-friendly activities, but they are also more emotionally attached to many environmental issues. This awareness of environmental consciousness generates a willingness to adopt eco-consumption behaviors among Chinese consumers (Chan, 1999; Chan & Lau, 2002; Chan & Wong, 2012).

Consumers with a high awareness of environmental consciousness are more likely to pay attention to and process environmental information, such as green campaigns and advertisements. Kim, Littrell, and Paff Ogle (1999) reported that since consumers who reside in developing countries exhibit environmental consciousness toward living conditions than developed countries, they are more likely to support apparel companies' sustainable development or missions moving toward social responsibility. Therefore, the following hypothesis was developed:

H1a: Consumers' environmental consciousness influences their intention toward eco-fashion consumption.

Environmental consciousness has been dramatically increasing its awareness among Chinese in recent years (Xue, Zhang, & Wang, 2014). Not only was this changing generated by the constant education campaigns from schools, NGOs, and popular media, but it has also been generated through real and painful conditions that the Chinese are suffering on a daily basis, such as breathing heavy smog which has blanketed many parts of China, especially in winter seasons. The Chinese have firmly realized that rapid economic development through mass production has come about at the cost of the domestic environment (Harris, 2006). However, the biggest micro-perspective upgrading among Chinese, especially those who are more educated and are living in the urban and suburb areas, is the recognition that their behaviors are causing environmental harm. The current study hypothesizes that Chinese consumers are more likely to identify themselves as green consumers and have a positive attitude toward eco-fashion product since their environmental consciousness has been increased. Oliver and Lee (2010) posited that based on self-image congruity theory, consumers' self-identity reflects their green ideology. According to Hunt and Vitell's (1986) theory of ethics, an individual's moral philosophy, such as environmental consciousness, forms their attitudes in situations relating to ethical issues. Therefore, the following two hypotheses were developed.

H1b: Consumers' environmental consciousness affects consumers' self-identity.
H1c: Consumers' environmental consciousness affects consumers' attitude toward eco-fashion consumption.
H1d: Consumers' environmental consciousness affects consumers' environmental apparel knowledge.

Environmental Apparel Knowledge

Consumer decision-making processes are associated with a constructive and cognitive route, and consumer knowledge reflects a cognitive aspect in one's mind. In consumer and behavioral studies, consumer knowledge has been examined as a critical factor or an antecedent in one's decision-making process. The theoretical definition of consumer knowledge consists of two dimensions: familiarity and product knowledge. Familiarity states the accumulated consumption experiences. Product knowledge refers to the sum of product information stored in one's memory (Philippe & Ngobo, 1999). Environmental knowledge, as one of the operational concepts built upon consumer knowledge, specifically emphasizes consumers' awareness of relevant environmental and ecological knowledge (Kang, Liu, & Kim, 2013). Therefore, the environmental apparel knowledge in the context of this study emphasizes consumers' awareness of relevant environmental and ecological knowledge and of apparel products and manufacturing processes.

Kim and Damhorst (1998) developed the environmental apparel knowledge scale which includes 11 statements concerning the processing of fibers, recycling, and biodegradability of apparel goods, chemical pollutants produced in the manufacture of apparel goods and regulations for clean air and water imposed on apparel and textile manufacturers. Chang (2012) posited that Chinese consumers have sufficient environmental apparel knowledge to perform more eco-consumption behaviors than those who do not possess such knowledge.

The findings from previous research regarding the relationship between environmental apparel knowledge, attitude, and eco-fashion consumption are inconclusive (Kang et al., 2013). For example, consumers who are more knowledgeable about environmental issues associated with apparel products are more likely to buy secondhand clothing or to purchase more classic styles for long-term usage. Hustvedt and Dickson (2009) concluded that apparel consumption behaviors were usually influenced by the awareness of environmental apparel knowledge. Those consumers who have a higher level of environmental apparel knowledge tend to evaluate environmental impacts of clothing products before buying. Kozar and Hiller Connell (2013) found that environmental apparel knowledge and consumer attitude toward apparel social responsibility issues were positively correlated. Consumers' environmental apparel knowledge was a significant predictor of eco-fashion purchasing behavior.

However, an earlier study of Kim and Damhorst (1998) found that environmental apparel knowledge had a limited relationship with one's environmental concern. It also has a weak effect on responsible apparel consumption behaviors. Based on participants' responses in Carrigan and Attalla (2001)'s study, it showed that consumers' apparel purchasing decisions were less likely to be influenced by apparel companies' environmental responsibility reputations. Noticing the results of environmental apparel knowledge's influence on apparel consumption behaviors and consumers' attitude, this study proposes the following two hypotheses:

H2a: Consumers' environmental apparel knowledge is positively associated with consumer attitude toward eco-fashion consumption.
H2b: Consumers' environmental apparel knowledge determines intention toward eco-fashion consumption.

Self-identity and Attitude

Self-identity is defined as the "relatively enduring characteristics that people describe themselves" (Sparks & Guthrie, 1998, p. 1396) and is synonymous with self-concept or self-perception. An individuals' self identity is influenced by personal motivations (such as self-esteem, self-enhancement, and self-understanding) and social interaction demands, expectations of others, and different roles that people perform (Whitmarsh & O'Neill, 2010). Self-identity is noted as a potential relevant determinant of one's behavior within sociology and psychological literature. In environmental science studies, researchers (Whitmarsh & O'Neill, 2010) have found that self-identity usually determines and affects pro-environmental attitudes and behaviors. Self-identity has been viewed as a predictor of behavioral intention (Granberg & Holmberg, 1990), which can be considered as a primary motive of pro-environmental behavior.

Based on the self-congruity theory (Sirgy, 1986), individuals who perceive themselves as green consumers are more likely to buy green products. Self-identity is found as a common motivational value which guides people's different eco-friendly behaviors, such as recycling (Manetti, Pierro, & Livi, 2004); using their own bags while shopping (Chan et al., 2008); buying organic products (Johe & Bhullar, 2016); and adopting the use of electric cars (Barbarossa, Beckmann, de Pelsmacker, Moons, & Gwozdz, 2015). Sparks and Shepherd (1992) applied the theory of planned behavior in examining the role of self-identity in green consumerism. They claimed that one's self-identity invokes some moral imperative or ethical-orientated behaviors.

Shaw, Shiu, and Clarke (2000) used the theory of planned behavior as a theoretical foundation to examine its predictability in purchasing fair trade groceries. They claimed that consumers who are driven by a sense of ethical obligation or green self-identification are more likely to buy fair trade products. A study conducted by Hustvedt and Dickson (2009) examined the effects of consumers' environmental attitude and self-identity on the intention of buying organic cotton apparel. The findings showed that consumers whose self-identity is environmental, organic, and socially responsible were more likely to have positive attitudes toward organic and sustainable agriculture. They are also more likely to buy locally made apparel items. Oliver and Lee (2010) found that consumers' self-identity has a direct effect on consumers' attitude toward green products. Barbarossa et al. (2015) proposed a self-identity-based eco-friendly intention conceptual model to evaluate the effects of self-identity and motivation's effect on electric car adoption intention.

The study posited that consumers' green self-identify and attitude are antecedents of eco-friendly conceptual intention and behavior, such as the adoption of eco-friendly cars. The model suggests that green self-identify and consumer attitude have a significant direct influence on the intention to adopt the use of electric cars. This study investigates the effects of consumers' identify in adopting high-involvement products, such as electric cars, while previous studies mainly focused on low-involvement products, such as fair trade groceries (Shaw & Shiu, 2003) and organic vegetables (Johe & Bhullar, 2016). Apparel products are often considered as high-involvement products which are more likely to be engaged with an individual's personal factors and self-values. Therefore, the following hypotheses were developed:

H3a: Consumers' self-identity is positively associated with consumer attitude toward eco-fashion consumption.
H3b: Consumers' self-identity directly influences consumers' intention toward eco-fashion consumption.

H4: Consumers' attitude directly influences consumers' intention toward eco-fashion consumption.

Gruen et al. (2006) applied the MOA theory in examining the customer-to-customer (C2C) e-communities' know-how exchanges. They found that in e-communities, the opportunity will be affected by individuals' level of attention and willingness of devotion to the C2C exchange. This finding can be leveraged to the context of this study since perceived opportunity in eco-fashion consumption can be influenced by one's attention and willingness of devotion in eco-fashion consumption. In addition, based on self-identify congruency theory, the willingness of devotion in eco-fashion consumption can be determined by one's green self-identity. Perceived ability in this study is defined as the perceived skills or proficiencies to engage in pro-environmental behaviors. It is also considered as one of the consequences of self-identity (Stets & Biga, 2003). Therefore, the following two hypotheses were developed:

H5a: Consumers' self-identity is positively associated with perceived opportunity.
H5b: Consumers' self-identity is positively associated with perceived ability.

Perceived Opportunity

Opportunity is the contextual or environmental mechanisms that are conducive to achieving desired goals or behavioral outcomes. Several variables, such as time, money availability, attention paid, and numbers of repetitions of availability, can enhance or impede the desired outcome. For example, Gruen et al. (2006) found that the level of opportunity determines customer-to-customer know-how exchange in Internet communities. MacInnis and Jaworski (1989) claimed that using some cues in ads, such as humor, celebrity, and incongruent stimuli, will increase opportunities for consumers to process brand information.

Opportunity in this research context is defined as the availability of resources which consumers can find eco-fashion products. Chan and Lau (2002) mentioned that Chinese consumers are very familiar with green consumer goods and have a more emotional attachment toward eco-friendly activities. Although there is a relatively limited availability of green products compared to Western countries, Chinese consumers still express their green concern freely as well as their willingness to acquire those green products. Chan (2000) revealed that even though Chinese consumers have a significant demand of green purchase intention, their actual purchase behavior is relatively insignificant. This low correspondence can be explained by the unavailability of green products in China. With increasing affluence derived from rapid economic growth in the past ten years, the fashion market has been abundantly developed. This development has triggered more diversified and niche fashion markets in China, such as secondhand, renting fashion stores, and handmade clothes. The following hypothesis was developed:

H6: The opportunity of shopping for eco-fashion influences Chinese consumers' intention toward eco-fashion consumption.

Perceived Ability

Ability implies some necessary resources such as money, intelligence, and knowledge that individuals have to make an outcome happen (Hoyer & MacInnis, 1997). We refer to ability under the MOA theory as consumers' skill or proficiencies in environmental behaviors. Cumming and Schwab (1973) mentioned that without ability and opportunity, motivation alone cannot lead to behaviors. Therefore, ability plays one of the complementary roles in influencing behaviors. Perceived ability in this study is reflected by being able to read the labeling on product packages to verify if the contents are environmentally friendly, to reduce water usage, and to save electricity. Previous studies (Chan & Lau, 2002; Kang et al., 2013; Ramasamy & Yeung, 2009) have found that Chinese consumers have become aware of the content of apparel products they buy. Some well-educated consumers even investigate whether the fashion supply chain they will shop at is ethical and transparent (Chan & Wong, 2012). Hence, the following hypothesis was developed:

H7: Perceived ability influences Chinese consumers' intention toward eco-fashion consumption.

In the process of developing the theoretical model in this study, researchers found that many MOA theory-related behavioral studies focused on examining motivation, opportunity, and ability's individual effects on behaviors. For example, MacInnis et al. (1991) identified internal and external factors which enhanced processing motivation, opportunity, and ability in brand information from advertisements. Gruen et al. (2006) identified individual effects of motivational variables, opportunity, and ability in C2C know-how exchanges. Most studies claim that opportunity and ability either directly determine or have moderating effects on individuals' behaviors. None of the current research has explored the relationship between opportunity and ability under the MOA theory. Chan and Lau (2002) mentioned that inadequate availability of eco-fashion products (lack of opportunity) may discourage Chinese consumers' eco-fashion consumption, even though they have the ability to perform relevant pro-environmental behaviors (Chan & Lau, 2002). Therefore, providing eco-fashion products to acquire opportunity (providing opportunity) might trigger people's perception of the ease of performing eco-fashion consumption (the perceived ability). Thus, the following hypothesis was developed:

H8: Consumers' perceived opportunity influences perceived ability.

Research Method

Sample and Data Collection

Data were collected in China. A snowball convenience sampling strategy was utilized. An online survey was set up on a Chinese online market research Web site. Participants were recruited through an email survey link on the social media app, *WeChat*. Two researchers sent an invitation with the survey link to their contacts on *WeChat* and also requested their contacts to share the survey with their contacts. Data cleaning yielded 401 usable samples out of 410 returned responses. The average age of the sample was 32; the age ranged from 18 to 68 years. Sixty-six percent were women. The participants were well educated with 80.8% indicating that they have at least a college education. More than 50% of them reported an annual household income of CNY100,000 (US$15,401) or above. See Table 1.

Instrument Development

Multi-item scales were used to evaluate seven variables, utilizing a 5-point Likert scale with "1 = strongly disagree to 5 = strongly agree." All of the measurements were adopted from previous studies. Environmental consciousness was measured

Table 1 Demographic summary of participants (n = 401)

Consumer profile	Frequency	Percentage
Gender		
Male	136	33.9
Female	265	66.1
Age		
18–30	206	51.4
31–40	150	37.4
41–68	45	11.2
Highest education		
High school graduate or lower	31	7.7
Associate degree	46	11.5
College graduate	216	53.9
Master/MBA or higher	108	26.9
Annual household income		
Less than CNY39,999 (Less than US$6,200)	33	8.2
CNY40,000–CNY69,999 (US$6,201–US$10,770)	77	19.2
CNY70,000–CNY99,999 (US$10,771.00–US$15,400)	83	20.7
More than CNY100,000 (More than $15,401)	208	51.9

by three items adopted from Gam (2011). Four items modified from Kim and Damhorst (1999) were utilized to measure environmental apparel knowledge, and a three-item scale for self-identify was adopted from Hustvedt and Dickson (2009). The semantic differential scale containing four groups of semantic differential adjectives was applied to measure attitude. In addition, perceived opportunity was measured by three items adopted and modified from Zhang (2014). Perceived ability was measured by three items developed by researchers for the current study. Finally, intention toward eco-fashion consumption was measured by a three-item scale modified from Chan and Wong (2012).

Data Analysis

SPSS 20.0 and Stata were used to perform statistical analysis. Reliability and validity were both examined in this study. A two-step approach was adopted (Anderson & Gerbing, 1988). Confirmatory factor analysis (CFA) was first conducted to find a better model for the measurement of each latent variable (Jackson, Gillaspy, & Purc-Stephenson, 2009). Then structural equation modeling (SEM) was employed to test the proposed hypotheses. The goodness-of-fit (GFI) indices used in this study include the conventional cut off ≥ 0.90 for acceptable fit for CFI and TIL, and ≥ 0.95 for good fit (Hu & Bentler, 1999). RMSEA and SRMR values between 0.05 and 0.08 represent acceptable fit, and values <0.05 indicate good fit (Kline, 2010; Hu & Bentler, 1999).

Results

Measurement Model

An exploratory factor analysis was performed first to examine the factor structure using a principal component analysis with varimax rotation. As expected, the screen plot examination confirmed the seven factors. All factor loadings ranged from 0.578 to 0.954 were satisfactory. Then a confirmatory factor analysis was conducted to test the measurement model using Stata. The result of the measurement model, including seven latent constructs with a total of 23 items, exhibited an excellent model fit ($\chi^2_{(df=209)} = 381.243$, $p = 0.000$, $\chi^2/df = 1.82$; RMSEA = 0.045; CFI = 0.962; TLI = 0.954; SRMR = 0.041)

Next, construct validity was confirmed by assessing convergent validity and discriminant validity. All CFA loadings were higher than 0.5, which provided evidence for convergent validity (Kline, 2010); and the average variance extracted (AVE) for each construct was greater than 0.5 (Anderson & Gerbing, 1988), suggesting that each construct is well represented by its own indicators. In addition, discriminant validity between constructs was also established because all AVEs ranging from 0.510 to 0.679 exceeded squared correlations between the constructs, ranging from 0.008 to 0.286 (Fornell & Larcker, 1981). Cronbach's alpha estimates of all constructs ranged from 0.755 to 0.885, which exceeded the recommended 0.70 cut off point (Cortina, 1993), confirming the reliability of measurement. Tables 2 and 3 represent the results regarding convergent and discriminant validities and the reliability of the instruments.

Structural Model and Hypotheses Test

Structural equation modeling (SEM) was employed to examine the influence of environmental consciousness, environmental apparel knowledge, attitude, opportunity, ability, and self-identity on Chinese consumers' intention to buy eco-fashion products. The statistic results revealed an excellent model fit ($\chi^2_{(df=210)} = 396.48$, $p < 0.000$, $\chi^2/df = 1.888$; RMSEA = 0.047; CFI = 0.959; TLI = 0.951; SRMR = 0.049).

The outputs demonstrated that environmental consciousness was positively associated with self-identity ($\beta = 0.541$, $p < 0.000$), environmental apparel knowledge ($\beta = 0.530$, $p < 0.000$), attitude ($\beta = 0.236$, $p < 0.001$), and eco-fashion consumption ($\beta = 0.229$, $p < 0.000$). Therefore, H1a, H1b, H1c, and H1d were all supported. The results indicate that Chinese consumers with higher environmental consciousness tend to have more environmental apparel knowledge, are more likely identify themselves as environmental consumers, feel positive toward purchasing and are more likely to purchase eco-fashion products. In addition, both environmental apparel knowledge ($\beta = 0.199$, $p < 0.001$) and self-identity ($\beta = 0.214$, $p < 0.001$) were found to be positively related to attitude, therefore, H2a and H3a were supported. Self-identity was also found to significantly impact perceived opportunity to sustainable consumption ($\beta = 0.503$, $p < 0.000$) and perceived ability to conduct sustainable consumption ($\beta = 0.403$, $p < 0.000$); therefore, H5a and H15b were supported. Furthermore, significantly positive relationships were also exhibited between eco-fashion consumption and self-identity ($\beta = 0.168$, $p < 0.046$), environmental apparel knowledge ($\beta = 0.145$, $p < 0.015$), perceived opportunity ($\beta = 0.165$, $p < 0.012$), and attitude ($\beta = 0.165$, $p < 0.01$). Therefore, H2b, H3b, H4, and H6 were all supported. There was no significant relationship between ability and eco-fashion consumption ($\beta = 0.152$, $p > 0.05$). Therefore, H7 was not supported. Additionally, the significant relationship between opportunity and ability was also confirmed ($\beta = 0.362$, $p < 0.000$), so H8 was supported. Figure 1 summarizes the hypothesis test results.

Table 2 Measurement model results

Constructs/indicators	CFA loading	Cronbach's α	AVE
Environmental consciousness		0.856	0.650
It is important to me that we try to protect our environment for our future generations	0.838		
It would mean a lot to me if I could contribute to protecting the environment	0.828		
The environment is one of the most important issues facing the world today	0.781		
Environmental apparel knowledge		0.848	0.613
Chemical pollutants are produced during manufacturing of synthetic or manufactured fibers such as polyester	0.713		
Dyeing and finishing processes use a lot of water	0.855		
Special finishes on fabrics may create problems for recycling	0.691		
Phosphate-containing detergents can be a source of water pollution	0.808		
Self-identity		0.786	0.569
I think of myself as an "environmental consumer"	0.818		
I think of myself as someone who is concerned about environmental issues	0.718		
I think of myself as an "organic consumer"	0.718		
Attitudes (I think that purchase environmentally friendly apparel products is…)		0.885	0.679
– Bad: Good	0.800		
– Foolish: Wise	0.917		
– Unsatisfying: Satisfying	0.780		
– Unfavorable: Favorable	0.776		
Opportunity		0.783	0.638
I know where I can find durable fashion product in market	0.823		
I know where I can get organic and environmentally friendly apparel product in the market	0.829		
I know where I can rent clothes	0.584		
Ability		0.755	0.510
I usually read the label on product packages to check if the contents are environmentally friendly	0.717		
I always minimize water use to help preserve fresh water	0.664		
I have the habit of reducing or optimizing my use of electrical devices to save electricity	0.786		
Eco-fashion consumption		0.807	0.566
I will buy clothing that is durable in the future	0.658		
I will buy clothing that is safe to the environment	0.824		
I would go out of my way to buy organic clothing	0.812		

Table 3 Discriminant and convergent validity of constructs

	1	2	3	4	5	6	7
1. Environmental consciousness	0.650						
2. Environmental apparel knowledge	0.008	0.613					
3. Self-identity	0.176	0.044	0.569				
4. Attitude	0.148	0.024	0.140	0.679			
5. Opportunity	0.192	0.111	0.157	0.168	0.638		
6. Ability	0.160	0.146	0.140	0.146	0.222	0.510	
7. Eco-fashion consumption	0.102	0.184	0.106	0.094	0.192	0.286	0.566

Note The numbers below diagonal are the squared correlation coefficient between the variables. The numbers in diagonal are the average variance extracted by each variable

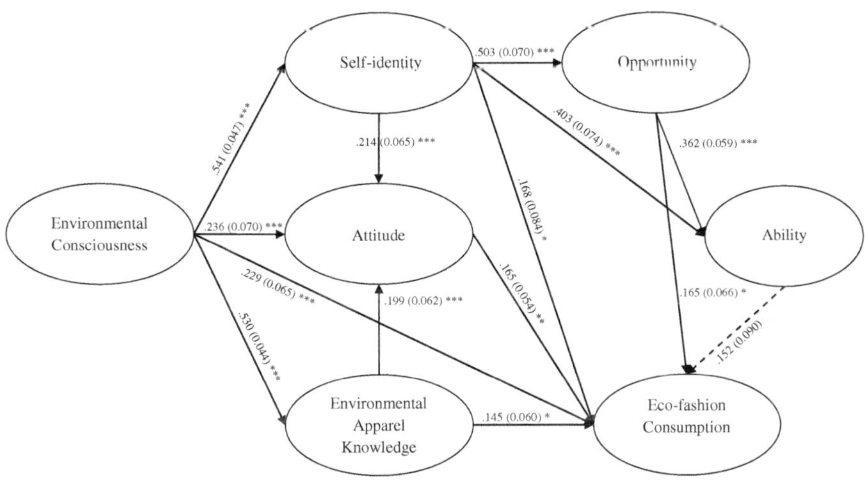

Fig. 1 Model of eco-fashion consumption intention (***$p < 0.001$, **$p < 0.01$, *$p < 0.05$; dotted line indicates insignificant path; standardized estimates with standard errors in parentheses)

Discussion

Due to the fast-growing environmental consciousness and knowledge as well as the increasing demand toward eco-friendly products among Chinese consumers, decision-makers and marketers are calling for more and further studies on moral and ethical motives of Chinese consumers' eco consumption behaviors. Not only does this present study address a certain part of this issue, but it also supports and extends the findings from previous studies through offering three contributions. First of all, many previous eco-consumerism studies applied Theory of Planned

Behavior (TPB) in learning eco-consumerism (Sparks & Shepherd, 1992; Robinson & Smith, 2002; Johe & Bhullar, 2016).

This research proposes a conceptual model based on Motivation-Opportunity-Ability (MOA) theory to examine Chinese eco-fashion consumption behaviors. This model emphasizes assessing the effects of motivational variables, such as self-identity, attitude, environmental consciousness and knowledge along with investigating the influences of opportunity and ability on eco-consumption behavior. The proposed conceptual model with MOA theory as a theoretical foundation has extended the scholarly scheme for understanding eco-consumption behavior. As predicted in the model, opportunity directly influences Chinese consumers' eco-fashion consumption intention. This result reveals that Chinese consumers have the intention to shop or rent for eco-fashion if they know where they can acquire them. Most tier-one and tier-two cities in China are still under transformational action processes toward sustainability, thus, adding opportunity as a determinate variable becomes critical. This is due to the fact that Chinese consumers' eco-fashion consumption behavior may be deterred or encouraged by what is being offered in the market. Ability shows an insignificant effect on eco-fashion consumption. This may suggest that ability toward performing pro-environmental behaviors may still be incompatible with or insufficiently correlated to eco-fashion consumption.

Secondly, this study explored whether motivational antecedents, such as Chinese consumers' man–nature-oriented environmental consciousness and attitude, the level of environmental knowledge, and green self-identity, influence their eco-fashion consumption. The results imply that environmental concerns and sufficient environment apparel knowledge motivate Chinese consumers to purchase eco-fashion products. Additionally, Chinese consumers who identify themselves as socially and environmentally responsible consumers have relatively positive attitudes toward eco-fashion products. Those who perceive related opportunity are also more likely to purchase eco-fashion items. The transformation of consumers' consumption attitude, motivation, and consumption patterns offers eco-fashion businesses the greatest opportunity to grow in China. Providing more organic and environmentally harmless fashion items has the potential to allow market and product development direction to respond to the transformation of Chinese consumers' consumption practices.

Eco- and sustainable consumption behaviors have been studied since the late 1970s. Most studies were conducted in Western countries, especially in Europe. Only in the past two decades has more attention been given to the understanding of eco-consumerism and pro-environmental behavior among Chinese, as well as how the traditional culture affects their pro-environmental behaviors (Harris, 2006; Chan et al., 2008). Subsequently, the sample from this study, a group of representative Chinese consumers who live in a tier-one city, was invited to measure the effects of antecedents of eco-shopping behaviors and then to predict their possible actions in the future.

The results of the structural equation modeling show that environmental consciousness is a critical antecedent in this proposed theoretical model. It has a significant effect on a consumer's self-identity, attitude and determines individuals' environmental apparel knowledge levels. Environmental consciousness directly influences Chinese consumers' purchase intention toward eco-fashion. These findings support the results from previous studies (e.g., Schlegelmilch et al., 1996; Chan & Wong, 2012; Kozar & Hiller Connell, 2013; Lin & Chang, 2012).

Implications

The eco-fashion market is still an emerging market in China, but the rising sophistication of Chinese consumers is allowing for a trend to shop for eco-products. Based on the results of this study, implications can be provided. First of all, researchers of this study believe that a general environmental consciousness is still an extremely important determinant in eco fashion consumption among Chinese consumers. The proposed conceptual model reveals that environmental consciousness builds up the linkage with several motivational factors, such as attitude, knowledge, and self-identity. Translating this information into a marketing perspective, due to environmental consciousness' direct influence on one's green self-identification, environmental apparel knowledge, attitude, and eco-fashion consumption, fashion companies have the responsibility of educating their consumers by providing relevant environmental knowledge to increase their awareness of environmental issues and their responsibility of protecting the environment to reach the goal of environmental sustainability. Researchers of this study insist on claiming that fashion companies should apply a business model which involves social and environmentally responsible practices. At the same time, creating different campaigns to enhance environmental consciousness among Chinese consumers will allow for an effective response to existing needs and potential future needs due to the upgraded Chinese consumers' consumption patterns. For example, hosting environmental consciousness workshops or inviting parents and their kids to participate in pro-environmental activities would not only provide the needed environmental education to consumers but would also potentially trigger new and/or more consumer needs toward eco-fashion products and relevant behaviors.

Harris (2006) conducted a study which provided a comprehensive understanding of Chinese environmental perspectives and behaviors. This study did not specify a particular environmental behavior being performed among Chinese consumers, but it provided for a macroscopic snapshot of how Chinese consumers view themselves in their state of environmental consciousness and this constructs a theoretical outline that further study can build upon to explore more particularistic actions. In the study, Harris pointed out the importance of creating a supportive and open media environment for promoting pro-environmental behavior. Indeed, the current Chinese government is promoting sustainability and sustainable development all over the country through a myriad of different environmentally conscious development plans. We can, therefore, predict that the diffusion of eco-fashion consumption behavior will be effectively spread out through Chinese society due to the consistency with the government's current sustainability goals. The vigorous potential market and the supportive political environment in China are providing great business opportunities for fashion companies who want to expand their eco-fashion lines and products. In addition, for marketers and fashion companies, it would be more effective to promote eco-fashion products through content marketing. For example, they could employ educational storytelling strategies: Tell consumers why they need to protect the environment and what possible positive results could be brought about by adopting these practices.

Even though the young generations, particularly Generation Y (born from 1980 to 1995) and Z (born from 1995), are performing with a relatively high self-awareness factor and individualism, they still possess traditional Chinese collectivism and cultural orientation rooted inside of them since they were raised by boomers (born from 1946 to 1960) and Generation X (born from 1961 to 1979 whose lifestyles and personalities were heavily influenced by the Chinese macro-social/political environment in 1950s–1980s. This means that the dominant shopping powers in China's current society (Gen X, Y, and Z) are still greatly influenced by collective selections. In another word, peers and community influences still play a major role in their decision-making process. In addition to providing continuing education regarding environmental knowledge to consumers, one of the most critical strategies for stimulating more eco-fashion consumption and developing a sustainable eco-environment is to place socially responsible pressure on social elites, such as celebrities, opinion leaders, decision-makers, pop stars, etc., due to the fact that consumers' decisions are often guided by those authorities. According to the diffusion of innovation theory, once the innovation of eco-fashion consumption is in place and has reached 20% of the population (innovators and early adopters), further innovation will rapidly occur involving the rest of the population (the majority, late adopters, and the laggards).

Another implication for the fashion industry and markets is that the current conceptual model shows that eco-fashion consumption can be achieved through several internal and external factors which companies may employ to trigger customers' needs. Some internal factors, such as consumers' environmental consciousness, self-identity, knowledge, and attitude, positively and directly influence eco-fashion consumption. Therefore, theoretically speaking, eco-fashion consumption may be achieved once an individual's environmental cognitive level, determined by those internal factors, reaches a certain point. A market and its consumers have a push and pull relationship. In another word, a market influences consumers by creating different potential needs which consumers are initially not aware of. Meanwhile, consumers also have power over markets due to the fact that consumer preferences vary and can be fast changing. So for fashion companies, it is important to create various and appealing campaigns to promote eco-fashion consumption, while at the same time facilitating sustainable development in the fashion industry to maintain their potential needs.

Conclusion

This study investigated motivational factors, including environmental consciousness, self-identity, attitude, and environmental apparel knowledge, and two external factors, including opportunity and ability's effect on eco-fashion consumption. Results show that with the exception of ability, all motivational factors and opportunity significantly and directly influence eco-fashion consumption. Participants in this study have sufficient ability in performing pro-environmental behaviors, such as reading the labeling on product packages to check if the contents are environmentally friendly, minimizing water use to help preserve fresh water and having the habit of reducing or optimizing the use of electrical devices to save electricity. Pro-environmental behavioral ability does not directly lead Chinese consumers to perform eco-fashion consumption behaviors. It is not until Chinese consumers are aware of the availability of eco-fashion products or they have solid information about where to acquire these products that eco-fashion consumption is fully adopted.

It is interesting to note that pro-environmental behavioral ability did not have a significant effect on eco-fashion consumption in either direct or indirect way. This may suggest that eco-fashion consumption is still a new shopping pattern among Chinese consumers. The ability to perform pro-environmental behaviors is compatible with eco-fashion consumption or is insufficiently linked to buying clothing that is durable and/or organic and/or safe to the environment.

Limitations and Suggestions for Further Study

There are several limitations that should be addressed in future research. Even though the sample size ($n = 401$) of this study was sufficient for conducting structural equation modeling as well as other statistical analysis methods, a convenience sampling, instead of a random selection, was used through an online survey system. Participants were recruited through email survey links and a social media app. All participants are current residents living in tier-one cities. This may result in a homogenous sample group. For further studies, recruiting random samples from cites in different tiers, such as consumers from less developed regions, may provide for a more comprehensive understanding of how Chinese eco-consumerism, as a whole, optimally functions. The boundless potential possibilities and diversity in the Chinese eco-fashion market are still insufficiently studied. A further in-depth study into consumption behaviors among cities in different tiers in China may prove to be valuable. Another limitation of this study is the dependent variable that was employed. Only eco-fashion consumption intention was measured in the proposed conceptual model. Consumers' actual eco-fashion consumption behaviors may also be tested, along with internal and external factors to increase the external validity of this study.

Further studies may benefit by examining which moral principle is employed, teleology or deontology, which Chinese consumers use when they are facing ethical situations. Chan et al. (2008) found that Chinese consumers were highly influenced by teleological considerations when engaged in moral obligations. For example, in their study, one of the Chinese consumer's pro-environmental behaviors is that they bring their own shopping bags when visiting markets and this behavior was significantly influenced by teleological evaluation, rather than ethical judgment. However, the cognitive decision-making process of deciding to bring consumers own shopping bags might be different from shopping for an eco-fashion product. Therefore, researchers who are interested in this topic and are using collectivism-dominated consumer cohorts as target research groups may want to examine which moral principle would better describe eco-fashion consumption behaviors.

References

Anderson, J. C., & Gerbing, D. W. (1988). Structural equation modeling in practice: A review and recommended two-step approach. *Psychooigical Bulletin, 103*(3), 411–423.

Argote, L., McEvily, B., & Reagans, R. (2003). Managing knowledge in organizations: An integrative framework and review of emerging rhemes. *Management Science, 49*(4), 571–582.

Barbarossa, C., Beckmann, S. C., de Pelsmacker, P., Moons, I., & Gwozdz, W. (2015). A self-identity based model of electric car adoption intention: A cross-cultural comparative study. *Journal of Environmental Psychology, 42,* 149–160.

Brodwin, D. (March 15, 2015). Unsustainable America. Retrieved from http://www.usnews.com/opinion/economic-intelligence/2015/03/23/china-and-india-ready-to-lead-in-consumer-sustainability.

Carrigan, M., & Attalla, A. (2001). The myth of the ethical consumer—do ethics matter in purchase behaviour? *Journal of Consumer Marketing, 18*(7), 560–578.

Chan, R. Y. K. (1999). Environmental attitudes and behavior of consumers in China: Survey findings and implications. *Journal of International Consumer Marketing, 11*(4), 25–51.

Chan, R. Y. K. (2000). An emerging green market in China: Myth or reality? *Business Horizons, 2,* 55–60.

Chan, R. Y. K., & Lau, L. B. Y. (2002). Explaining green purchasing behavior: A cross-cultural study on American and Chinese consumers. *Journal of International Consumer Marketing, 14*(2/3), 9–41.

Chan, T. Y., & Wong, C. W. Y. (2012). The consumption side of sustainable fashion supply chain Understanding fashion consumer eco-fashion consumption decision. *Journal of Fashion Marketing and Management, 16*(2), 193–215.

Chan, R. Y. K., Wong, Y. H., & Leung, T. K. P. (2008). Applying ethical concepts to the study of "green" consumer behavior: An analysis of Chinese consumers' intentions to bring their own shopping bags. *Journal of Business Ethics, 79*(4), 469–481.

Chang, C.-T. (2012). Are guilt appeals a panacea in green advertising? The right formula of issue proximity and enviornmental consciousness. *International Journal of Advertising, 31*(4), 741–771.

Claudio, L. (2007). Waste couture: Environmental impact of the clothing industry. *Environmental Health Perspectives, 115*(9), 449–454.

Cordano, M., Welcomer, S. A., & Scherer, R. F. (2003). An analysis of the predictive validity of the new ecological paradigm scale. *Journal of Environmental Education, 34*(3), 22–28.

Cortina, J. M. (1993). What is coefficient alpha? An examination of theory and applications. *Journal of Applied Psychology, 78*(1), 98–104.

Cummings, L. L., & Schwab, D. P. (1973). *Performance in organizations: Determinants & Appraisal*. Glenview, III: Scott, Foresman.

Dunlap, R., Van Liere, K., Mertig, A., & Jones, R. E. (2000). Measuring endorsement of the new ecological paradigm: A revised NEP scale. *Journal of Social Issues, 56,* 425–442.

Dunlap, R. E., & Jones, R. E. (2002). *Environmental concern: Conceptual and measurement issues*. CT: Greenwood Press.

Fornell, C., & Larcker, D. F. (1981). Evaluating structural equation models with unobservable variables and measurement error. *Journal of Marketing Research, 18*(1), 39–50.

Gam, H. J. (2011). Are fashion-conscious consumers more likely to adopt eco-friendly clothing? *Journal of Fashion Marketing and Management, 15*(2), 78–193.

Graberg, D., & Holmberg, S. (1990). The intention-behavior relationship among U.S. and Swedish Voters. *Social Psychology Quarterly, 53*(1), 44–54.

Gruen, T. W., Osmonbekov, T., & Czaplewski, A. J. (2006). eWOM: The impact of customer-to-customer online know-how exchange on customer value and loyalty. *Journal of Business Research, 59*(4), 449–456.

Harris, P. G. (2006). Environmental perspectives and behavior in China. *Environment and behavior, 38*(1), 5–21.

Hiller Connell, K. Y., & Kozar, J. M. (2012). Social normative influence: An exploratory study investigating its effectiveness in increasing engagement in sustainable apparel-purchasing behaviors. *Journal of Global Fashion Market, 3*(4), 172–179.

Hoyer, W. D., & MacInnis, D. J. (1997). *Consumer behavior*. Boston: Houghton Mifflin Company.

Hu, L.-T., & Bentler, P. M. (1999). Cutoff criteria for fit indexes in covariance structure analysis: Conventional criteria versus new alternatives. *Structural Equation Modeling, 6*(1), 1–31.

Hunt, S. D., & Vitell, S. (1986). A general theory of marketing ethics. *Journal of Macromarketing, 6*(1), 5–16.

Hustvedt, G., & Dickson, M. A. (2009). Consumer likelihood of purchasing organic cotton apparel: Influences of attitudes and self-identity. *Journal of Fashion Marketing and Management, 13*(1), 49–65.

Jackson, D. L., Gillaspy, J. A. J., & Purc-Stephenson, R. (2009). Reporting practices in confirmatory factor analysis: An overview and some recommendations. *Psychological Methods, 14*(1), 6–23.

Joergens, C. (2006). Ethical fashion: Myth or future trend? *Journal of Fashion Marketing and Management, 10,* 360–371.

Johe, M. H., & Bhullar, N. (2016). To buy or not to buy: The roles of self-identity, attitudes, perceived behavioral control and norms in organic consumerism. *Ecological Economics, 128,* 99–105.

Kang, J., Liu, C., & Kim, S. H. (2013). Environmentally sustainable textile and apparel consumption: The role of consumer knowledge, perceived consumer effectiveness and perceived personal relevance. *International Journal of Consumer Studies, 37,* 442–452.

Kim, H.-S., & Damhorst, M. L. (1998). Environmental concern and apparel consumption. *Clothing and Textiles Research Journal, 16*(3), 126–133.

Kim, H.-S., & Damhorst, M. L. (1999). Environmental attitude and commitment in relation to ad message credibility. *Journal of Fashion Marketing and Management, 3*(1), 18–30.

Kim, S., Littrell, M. A., & Paff Ogle, J. L. (1999). The relative important of social responsibility as a predictor of purchase intentions for clothing. *Journal of Fashion Marketing and Management, 3*(3), 207–218.

Kline, R. B. (2010). *Principles and practice of structural equation modeling* (3rd ed.). New York: The Guilford Press.

Kluckhohn, F. R., & Strodtbeck, F. L. (1961). *Variations in value orientations*. Evanston, Illinois: Row, Peterson.

Kozar, J. M., & Hiller Connell, K. Y. (2013). Socially and environmentally responsible apparel consumption: Knowledge, attitudes, and behaviors. *Social Responsibility Journal, 9*(2). 1747-1117.

Lin, Y.-C., & Chang, C. A. (2012). Double standard: The role of environmental consciousness in green product usage. *Journal of Marketing, 76,* 125–134.

MacInnis, D. J., & Jaworski, B. (1989). Information processing from advertisements: Toward an integrative framework. *Journal of Marketing, 53,* 1–23.

MacInnis, D. J., Moorman, C., & Jaworski, B. J. (1991). Enhancing and measuring consumers' motivation, opportunity, and ability to process brand information from ads. *Journal of Marketing, 55*(4), 32–53.

Mannetti, L., Pierro, A., & Livi, S. (2004). Recycling: Planned and self-expressive behaviour. *Journal of Environmental Psychology, 24,* 227–236.

Niinimäki, K. (2010). Eco-clothing, consumer identity, and ideology. *Journal of Sustainable Development, 18*(3), 150–162.

Oliver, J. D., & Lee, S. H. (2010). Hybrid car purchase intentions: A cross-cultural analysis. *Journal of Consumer Marketing, 27*(2), 96–103.

Ochoa, L. M. C. (2011). Will 'eco-fashion' take off? A survey of potential customers of organic cotton clothes in London. Retrieved from http://publicaciones.eafit.edu.co/index.php/administer/article/view/188#.U5JZMdhmSbU.

Park, C. W., & Mittal, B. (1985). A theory of involvement in consumer behavior: Problems and issues. In J. N. Seth (Ed.), *Research in consumer behavior* (Vol. 1, pp. 201–231). Greenwich: JAI Press.

Philippe, A., & Ngobo, P. V. (1999). Assessment of consumer knowledge and its consequences: A multi-component approach. *Advances in Consumer Research, 26,* 569–575.

Ramasamy, B., & Yeung, M. (2009). Chinese consumers' perception of corporate social responsibility (CSR). *Journal of Business Ethics, 88*(1), 119–132.

Robinson, R., & Smith, C. (2002). Psychosocial and demographic variables associated with consumer intention to purchase sustainably produced foods as defined by the Midwest food alliance. *Journal of Nutrition Education and Behavior, 34*(6), 316–325.

Schlegelmilch, B. B., Bohlen, G. M., & Diamantopoulos, A. (1996). The link between green purchasing decisions and measures of environmental consciousness. *European Journal of Marketing, 30*(5), 35–55. https://doi.org/10.1108/03090569610118740.

Siemsen, E., Roth, A. V., & Balasubramanian, S. (2008). How motivation, opportunity, and ability drive knowledge sharing: The constraining-factor model. *Journal of Operations Management, 26*(3), 426–445.

Stets, J. E., & Biga, C. F. (2003). Bringing identity theory into environmental sociology. *Sociological Theory, 21*(4), 398–423.

Sirgy, M. J. (1986). *Self-Congruity: Toward a Theory of Personality and Cybernetics.* Oraeger Publisher: Greenwood Publishing Group.

Shaw, D., Shiu, E., & Clarke, I. (2000). The contribution of ethical obligation and self-identity to the theory of planned behavior: An exploration of ethical consumers. *Journal of Marketing Management, 16*(8), 879–894.

Shaw, D., & Shiu, E. (2003). Ethics in consumer choice: A multivariate modeling approach. *European Journal of Marketing, 37*(10), 1485–1498.

Shen, B., Wang, Y., Lo, C. K. Y., & Shum, M. (2012). The impact of ethical fashion on consumer purchase behavior. *Journal of Fashion Marketing and Management, 16*(2), 234–245.

Sparks, P., & Guthrie, C. A. (1998). Self-identify and the theory of planned behavior: A useful addition or A unhelp artifice. *Journal of Applied Social Psychology, 28*(15), 1393–1410.

Sparks, P., & Shepherd, R. (1992). Self-identity and the theory of planned behavior: Assessing the role of identification with "green consumerism". *Social Psychology Quarterly, 55,* 388–399.

Whitmarsh, L., & O'Neill, S. (2010). Green identity, green living? The role of pro-environmental self-identity in determining consistency across diverse pro-environmental behaviors. *Journal of Environmental Psychology, 30*(4), 335–344.

Xiao, G., & Kim, J.-O. (2009). The investigation of Chinese consumer values, consumption values, life satisfaction, and consumption behaviors. *Psychology & Marketing, 26*(7), 610–624. https://doi.org/10.1002/mar.20291.

Xue, H. Y., Zhang, X. J., & Wang, Y. Q. (2014). Research on the disposal strategy of waste textiles. *Applied Mechanics and Materials, 522,* 817–820.

Yau, O. H. M. (1988). Chinese cultural values: Their dimensions and marketing implications. *European Journal of Marketing, 22*(5), 44–57.

Ye, Q. (2000). *Green marketing*. Wuhan: Hubei People's Publication.

Young, W., Hwang, K., McDonald, S., & Oates, C. J. (2010). Sustainable consumption: Green consumer behavior when purchasing products. *Sustainable Development, 18*(1), 20–31.

Zelezny, L. C., & Schultz, P. W. (2000). Promoting environmentalism. *Journal of Social Issues, 56*(3), 365–371.

Zhang, R. (2014). Sustainable apparel consumption: Scale development and validation (Unpublished doctoral dissertation). Oregon State University, Corvallis, OR.

Zheng, Y. (2010). Association analysis on pro-environmental behaviors and environmental consciousness in main cities of East Asia. *Behaviormetrika, 37*(1), 55–69.

Part IV
Fashion Branding in China

Licensing or not Licensing? A Pilot Study on Examining the Apparel Extension of Auto Brands Ford and Jeep in China Market

Chuanlan Liu, Yao Wei and Langchao Zhang

Abstract This study intends to shed some light on understanding how consumers evaluate and accept global brands' apparel extension in the world largest consumer market to provide practical directions for apparel manufactures to make licensing decisions. Specifically, this study applied the brand concept mapping (BCM) approach to evaluate an auto brand's fashion clothing extension. Snowball sampling was used to collect data in major cities in China. Sixty participants were recruited and divided into four groups to collect responses toward two auto brands and the two auto brands' apparel extensions, respectively. Consensus maps were generated based on individual maps. The comparison of BCMs showed that strong brand associations from the parent brand transferred well to the apparel extension, even when the "fit" between the parent brand and its apparel extension was poor. In the category of apparel, country of origin can be leveraged for manufacture licensees to compete for a market share. However, brand personality plays a more important role in consumers' acceptance of such a long-distance brand stretching. In addition, the level of individual consumer's brand engagement can be a good indicator for identifying target customer segment.

Keywords Licensing · Brand extension · Brand concept mapping (BCM)
Auto brand extension · Brand engagement

Introduction

Because of marketers' aggressive branding efforts, today's consumers are bombarded daily with numerous ads (Zhuang, Wang, Zhou, & Zhou, 2008). To avoid the high cost and difficulty of introducing a new product to the market, many firms

C. Liu (✉) · Y. Wei · L. Zhang
Louisiana State University, 145 Human Ecology Building, Baton Rouge, USA
e-mail: clliu@lsu.edu

Y. Xu et al. (eds.), *Chinese Consumers and the Fashion Market*, Springer Series in Fashion Business, https://doi.org/10.1007/978-981-10-8429-4_7

choose to take advantage of established brands to launch new products lines or categories (Aaker & Keller, 1990). One such approach is brand extension, which is defined as "using a current brand name to enter a completely different product class" (Aaker & Keller, 1990, p. 27). As stated by Morein (1975), leveraging a well-established brand name could reduce the risk of product failure significantly, as consumers are already familiar with the brand and have gained enough knowledge about it for decision making. Brand extension has become a marketing strategy used widely to maximize firm success by gaining a larger market share, at a significantly lower cost.

While brand extensions and licensing are used to be considered as two distinct branding strategies, extant literature suggests that licensing be treated as an "external" brand extension (Walsh, Rhenwrick, Williams, & Waldburger, 2014). Brand licensing refers to a process of creating and managing contracts for a brand owner (a licensor) to give another firm (a licensee) the rights to produce and sell a product with the brand on it (Jayachandran, Kaufman, Kumar, & Hewett, 2013). Many automobile and machinery brands have been involved actively in brand licensing. For example, both Jeep and Ferrari have launched an infant stroller line through licensing. Caterpillar, a brand of earthmoving equipment, has gained huge success by creating a line of CAT work boots and expanding to a whole casual/work footwear collection in markets in more than 150 countries. John Deere, another brand of machinery, now offers a selection of goods ranging from toys, strollers, and travel mugs to clothing, accessories, and iPhone cases.

The economic downturn in particular has provoked more automobile brands to seek licensing and brand extension opportunities. Meanwhile, Chinese apparel manufacturers have been working as contractors for many global brands developed in North America and Europe for the last three decades. With China's economic growth gradually going down, cost of production increasing, Chinese apparel industry has to upgrade and move more toward the downstream of apparel supply chain with more focus on the domestic consumer market. Brand is the core selection factor for consumers to make choices on apparel. Some of the manufactures take a shortcut for branding their products by obtaining a license from a global brand. Taking the advantages of high brand awareness of global brands and low consumer brand knowledge among Chinese consumers, many licensed global brands' fashion extension succeeded in China's fashion market. However, with the diffusion of information technology and globalization of Chinese consumers, the landscape of branding products through licensing has been changing. Just having a brand name or label is far from successfully attracting and maintaining customers. It is critical to understand how to develop products which can better leverage the paid brand license.

Consumers develop attitudes toward a certain brand based on their previous experiences with the brand (Priluck & Till, 2010). These experiences often form a network structure that consists of linked fragments of information, such as product packaging, slogan, and advertising. Hence, understanding attitudes toward a brand and brand extension requires identifying the underlying structure of brand associations. The existing techniques available for eliciting consumers' brand associations include both qualitative and quantitative approaches, such as collages, focus groups, and brand personality inventories (John, Loken, Kim, & Monga, 2006). However, these existing techniques lack the ability to show the structure of the association network. The "brand concept map" (BCM) is a relatively new method, which provides a map of the association networks that consumers form in their minds (John et al., 2006). It pinpoints the most important associations, identifies how consumers connect them to the brand, and reveals the structure of those interconnections. By obtaining a BCM from each consumer and then aggregating the individual data to produce a consensus map, a clear picture of how consumers perceive a brand and its extension can be obtained to facilitate brand decisions.

The fashion industry is readily associated with branding. Given the nature and competitiveness of fashion markets, many researchers have discussed the need to use symbolic cues to create distinctive brand images and associations in this particular industry (Carroll, 2009; Susan & Richard, 1998). Research has suggested that the key factors, in the success of brand extension, are the fit between the parent brand and the extension, and how well the brand equity of the parent brand transfers to the extension (Hem, De Chernatony, & Iversen, 2003; Völckner & Sattler, 2006). However, many successful fashion brand extensions seem to have a relatively weak fit, with respect to product features, e.g., Caterpillar (CAT) brand boots, Jeep brand apparel, and so forth. The transfer of favorable brand associations may play a more important role in the success of a new category. However, there has been little research that explores how favorable parent brand associations are transferred to extensions and affect consumers' acceptance of brand extensions. Applying BCM to brand extension studies in fashion categories will help us understand consumers' perceptions of apparel and fashion extensions. It will also help track how and which associations of the parent brand are transferred to facilitate the success of extensions.

Sprott, Czellar, and Spangenberg (2009) proposed the concept of brand engagement in self-concept (BESC), which is defined as, "an individual difference representing consumers' propensity to include important brands as part of how they view themselves" (p. 92). According to their research, a higher BESC level can lead to better brand recall, more brand awareness and greater favorable brand attitudes. Considering the effect of BESC on consumers' evaluations of brand extensions, it is necessary to examine its effect on consumers' formation of brand associations and their evaluations of brand extension.

Although BCM has already been introduced in general fashion branding research (Shin, 2011), it has not yet used to evaluate brand extension, particularly in fashion clothing categories. Our study intended to shed some light on understanding consumer evaluations of fashion clothing brand extension using the BCM approach to provide direction for manufactures to made licensing decision. Specifically, the objectives were to (1) identify consumer associations with Ford and Jeep automobiles and apparel using BCM; (2) categorize those associations; and (3) compare the brand associations and brand images of Ford and Jeep apparel in the Chinese market.

Ford and Jeep apparel were selected because they are both extensions of iconic American automobile brands. In 2002, an apparel line that targets male consumers from 18 to 45 years old was launched in China as an extension to the automobile brand of Jeep. In the last 10 years, Jeep apparel has experienced great success in China. Ford apparel is a relatively new brand compared to Jeep. Ford licensed its apparel and accessory lines in China in 2012. Ford menswear also targets male consumers in the same age range as Jeep, so comparisons between a new brand extension and a successful extension will be useful for business practitioners and researchers.

Literature Review

Brand Extension

Previous research has demonstrated that the "fit" between the parent brand and the extension and the perceived quality of the parent brand are crucial factors in the success of extensions (Bottomley & Holden, 2001). According to Völckner and Sattler (2006), approximately 15 determinants of the success of extensions were significant ($p < 0.10$) in at least one empirical study. Previous research has also found that the psychological process consumers experience in evaluating an extension plays an important role (Klink & Smith, 2001). According to Martinez (2011), two psychological paths have been used to explain consumers' decision process: categorical and piecemeal evaluation. As stated by Fiske (1982), "categorical evaluation" refers to the situation in which individuals evaluate a new instance by comparing the new information with that stored in memory. In contrast, "piecemeal evaluation" (Fiske, 1982; Martinez, 2011) is based on specific attributes (Bristol, 1996). These two evaluation processes are not exclusive, as consumers might activate a categorical evaluation process when they identify a good fit between the parent brand and extended product categories (Boush & Loken, 1991). Meanwhile, knowledge and attitudes associated with the parent brand are transferred to the extension (Martinez, 2011; Romeo, 1991). However, when the fit is poor, a piecemeal process is developed as consumers evaluate the extension by assessing its individual attributes (Fiske & Pavelchak, 1986). Hence, a two-stage process is used commonly as consumers go through the evaluation process. At the first stage, they try to transfer existing associations and attitudes to the extension to assist the evaluation; if unsuccessful, a second stage will be initiated in which an analytical evaluation is established (Fiske & Pavelchak, 1986; Martinez, 2011).

The decomposition of consumers' psychological processes has revealed the importance of brand associations in the success of brand extensions. Consumers' evaluation process could be seen as a brand association transfer process. Brands that hold strong, favorable, and unique brand associations that transfer well from the parent brand to the extension are more likely to extend successfully (Keller, 1993).

Particularly, when the fit is poor, emphasizing strong brand associations becomes more important, as the similar characteristics, personalities, and features may help shift consumers' attention from the poor fit to shared associations, and thus smooth the acceptance process. Leveraging strong, favorable, and unique brand associations could successfully transfer intangible components of the brand stored in the consumers' mind to the new products.

Brand Associations and Brand Association Categories

According to Dew and Kwon (2010), a consumer's perceptions about a brand exist in his/her mind as a brand association network that reflects the brand image. As stated by Aaker (1991), brand association refers to anything that is linked to the brand in a consumer's mind. It could be any opinion, though, impression or experience consumers hold about a particular brand (Aaker, 1991; Keller, 1993). These associations vary in form. They may be tangible or intangible, reflecting product-related, functional associations, or non-product-related, experiential associations (Shocker, Srivastava, & Ruekert, 1994). Aaker (1991) also discussed classification and dimensionality of brand associations and categorized 11 dimensions of brand associations. Korchia, 1999 (p. 15) examined the weakness of existing classifications further and proposed 15 categories of associations including the company, other organizations, brand personality, celebrities and events, typical users, typical usage situations, product category, price, communication, distribution, product-related attributes, functional benefits, experiential benefits, symbolic benefits, and attitudes. Compared to previous brand association dimensions, Korchia's 15 categories are more specific and better suited for the fashion industry (Shin, 2011).

Most existing research explains the formation and operation of brand association networks using human associative memory (HAM) theory derived from cognitive psychology (Anderson & Bower, 1980; French & Smith, 2013; Wyer, Budesheim, Lambert, & Swan, 1994). According to HAM theory, brand knowledge is stored in a consumer's mind as small pieces of information connected to each other to form a complicated brand information network (Anderson, 1983). The brand recall process is an "activation-spreading" process. When one information node of the brand is activated, it evokes another node linked to it and "spreads" to all the nodes in the network to complete the brand recall (Anderson, 1983). According to Keller (1993), associations are different in terms of how relevant they are to consumers, how well they are connected to the brand, and how unique they are compared to competitors' associations.

Measuring Brand Associations

Keller (2008) suggested that qualitative techniques, such as free association, can be used to identify possible associations. The free association method asks an individual or small group of respondents to describe what comes to mind when they think of a brand. Normally, it is used as a first step by which researchers or marketers determine a basic idea of the range of possible associations and the favorability, strength, and uniqueness of the associations identified. Keller (2008) also introduced other useful techniques, such as projective techniques and the Zaltman Metaphor Elicitation Technique (ZMET). Projective techniques (Levy, 1981, 1985) may involve asking respondents to complete sentences or interpret pictures about a brand to reveal their real feelings and thoughts on certain aspects of the brand. The method is largely oriented toward experience and feelings. It focuses primarily on brand users, and their decision processes and perspectives. The ZMET is a more complicated technique that uncovers the association networks in consumers' minds with a series of qualitative methods, such as taking photos, collecting pictures, drawing images, and giving interviews. Another example, perceptual mapping (Henderson, Iacobucci, & Calder, 1998), is a technique that makes pairwise comparisons of brand attributes. Respondents are provided with a set of brands and asked to compare all given attributes of every two brands.

Although the techniques discussed above elicit brand associations from different perspectives, they all have obvious weaknesses. Most of them cannot identify the relationships among associations. Some techniques, such as ZMET and perceptual mapping, are very labor-intensive.

Brand Concept Mapping (BCM)

BCM is a standardized brand mapping technique that identifies the structure of consumers' brand association networks by eliciting an aggregated BCM from a sample of individually-produced maps. Although the procedure for data collection and map aggregation sounds similar to the other mapping techniques described above, BCM is less labor-intensive, easier to administer and more accessible. According to John et al. (2006), BCM not only identifies important brand associations but also conveys the connections among associations. Further, it may be used on large samples or to compare consumer segments. Moreover, it does not require specially trained interviewers or complex analytical techniques.

BCM is a three-stage mapping technique. The first stage is elicitation in which salient associations are selected based on a set of criteria. The second stage is mapping, in which participants are asked to create a map. The third stage is aggregation, in which a consensus map is created based on aggregation rules. Table 1 shows the aggregation rules (John et al., 2006; Shin, 2011). Frequency counts are the main determinant of whether or not an association is included in the

Table 1 Aggregation rules for BCMs (John et al., 2006; Shin, 2011)

Steps	Measures	Rules
1. Select core brand associations	Frequency of mention Number of interconnections	Select brand association that is: • Included on at least 50% of maps • Included on 45–49% of maps if the number of connections for core associations we identified previously
2. Select first-order brand associations	Frequency of first-order mentions Ratio of first-order mention Type of interconnections	Select core brand association that is: • Have a ratio of first-order mentions to total mentions of at least 50% • Have more superordinate than subordinate interconnections
3. Select core brand association links	Frequencies for association links	Select core brand association links by • Finding infection point on frequency plot • Inflection point = target number • Including all association links that appear on or above the target number of maps
4. Select non-core brand association links	Frequencies for association links	Select non-core brand association links that are: • Linked to a core brand association • Linked on or above the target number of maps
5. Select number of connecting lines	Mean number of lines used per link	Select single, double, or triple lines for each brand association link by: • Determining the mean number of lines used per link • Rounding up or down to the next integer number (e.g., 2.3 = 2)

consensus map (John et al., 2006; Shin, 2011). Group testing is used to assess reliability and validity. As summarized by Shin (2011), the visual presentation of BCM is particularly helpful in detecting problematic associations that are inconsistent with a firm's brand positioning. The interconnections also represent the relationships between associations and consumers' logic chain clearly. Further, the use of a map allows marketers to keep tracking changes of association networks over time. These features can be very useful and powerful for brand extension evaluation. By mapping both the parent brand and the extension, marketers could obtain a clear idea of the brand image and key associations and what associations do and do not transfer well from the parent brand to the extension. With such crucial information in hand, marketers would be able to adjust the positioning of the extension, and related marketing campaigns and promotions accordingly.

Brand Engagement

Prior research has demonstrated that higher self-brand connection increases favorable brand attitude (Sprott et al., 2009; Escalas, 2004; Escalas & Bettman, 2005). Self-concept not only affects consumers' attitudes toward a specific brand but also tends to have an effect on their general engagement with brands. Realizing the fact that consumers vary in their tendency to incorporate important brands in their self-concept, Sprott et al. (2009) proposed the concept of brand engagement in self-concept (BESC). According to their research, a higher BESC level can lead to better brand recall, more brand awareness and greater favorable brand attitudes. Considering the effect of BESC on consumers' evaluations of brand extensions, it is necessary to examine its effect on consumers' formation of brand associations and evaluations of brand extension.

Methodology

Research Design and Procedure

This study attempted to use BCM in the evaluation of brand extensions in the fashion market; identify and categorize strong brand associations with the parent brands and extensions; investigate which associations did or did not transfer well from parent brands to extensions, and compare the two cases horizontally and vertically. The empirical study procedure included three stages: eliciting, individual mapping, and aggregation. The process began by identifying possible brand associations for each brand in a word bank to be used in the mapping stage. Next, interviews were conducted to obtain BCMs from the collected responses. Thereafter, aggregating individual maps were conducted to construct the consensus map for each brand and its brand extension following the adopted aggregation rules for BCMs (see Table 1). Then, three pairs of comparisons were conducted, including the Jeep auto brand and Jeep menswear extension, the Ford auto brand and Ford menswear extension, and Jeep menswear and Ford menswear.

In the eliciting stage, two sets of word banks (for cars and for menswear of the two brands) were established. For the automobile brands, online interviews were conducted with ten participants, using open-ended questions to gather information and identify salient brand associations for further mapping. General questions were asked, for example, "How do you like Jeep?" and "When thinking about Ford, what comes to mind?" Respondents were encouraged to talk about their thoughts and feelings about the two car brands, respectively. Words were also adopted from other sources, including official Web sites for each brand, social network trending topics mentioned, car forums mentioned, and online searches. Table 2 presented word bank for Jeep auto. Among collected responses, frequently mentioned brand associations were gathered as word bank for further data collection and mapping.

Table 2 Word bank for Jeep auto

High quality Well-made Durable Classic Comfortable Off-road Stylish Expedition Well-respected Sleek Outdoorsy Family Cool looking Unique Sporty Wild Edgy Reliable Exciting Expensive Simplistic Powerful Army Poor gas mileage SUV American Green Four-wheel drive Muddy Traction Always in style Popular Tough Adventurous Nice Young Famous Logo Dirty Cloth-top Tall Tow Dependable Cherokee Wrangler Military Active Chrysler Rugged Strong Confident Reputable Practical Affordable Luxurious Fun Timeless Sophisticated Manly Trip Camping Mountain Snow Interesting Big tires Modern Well-known

For menswear, a separate survey regarding favorite menswear brands and relevant brand associations was conducted in an undergraduate merchandising class. The students listed their favorite brands in the menswear segment and their reasoning, including relevant brand associations. Some of the words mentioned frequently were selected for building word banks for Jeep and Ford menswear. Other words were adopted from the word banks of Jeep and Ford cars. Table 3 presented word bank for Jeep menswear

After developing the word banks, two rounds of pretests were conducted. The first round was in English, and the second was in Chinese. The objectives of the pretests were to collect more salient brand associations; identify whether the words provided were relevant and representative; test whether the wording and layout were clear to the participants; test in-person interview and online interview result consistency, and test English/Chinese language result consistency. For the first round of pretest, ten participants were recruited to obtain Jeep car BCMs and four participants were recruited to obtain Jeep menswear BCMs. Modifications were made accordingly based on the results of the first round pretest, after which and the survey was translated into Chinese. Ten people were recruited for the second round pretest with six participants for the Jeep auto and four participants for Jeep menswear.

Table 3 Word bank for Jeep menswear

High quality Well-made Durable Classic Modern Fashion forward Stylish Trendy Good design Great fitting Attractive Cool looking Unique Sporty Outdoorsy Clean cut Expensive Simplistic Chic Versatile Tailored American Jeep car Sensible Mannered Every day wear Edgy Preppy Always in style Popular Well-respected Casual Tough Charming Nice Young Upper class Reputable Prestigious Comfortable Washes well Reliable Sustainable Practical Affordable Accessible Luxurious Fun Adaptable Timeless Confident Sophisticated Manly

Table 4 Word bank for Ford auto

High quality Well-made Durable Classic Rustic Pickup truck Comfortable Stable Conservative
Safe Old brand Basic Focus Well-respected For business use Reliable Simplistic Powerful Poor gas mileage Mature Down to earth Mainstream Cheap American Global brand Cost-effective Manly Tough Famous Logo Tall Big Dependable Confident Solid Roomy Low key For senior Not fun Mediocre Reputable Practical Affordable Well-known For family Steady Hi-tech Changan Ford Boring Second tire brand Collaboration with China Dull colors Detroit Lack of vigorousness Mondeo Kuga Drives well Slow start Crushproof

Table 5 Word bank for Ford menswear

High quality Well-made Durable Classic Modern Fashion forward Stylish Trendy
Good design Great fitting Attractive Basic styles Ford Car Clean cut Inexpensive Simplistic Chic Versatile Tailored American Sensible Mannered Every day wear Well-respected Casual Tough Charming Nice Young Reputable Prestigious Comfortable Washes well Reliable Sustainable Practical Affordable Accessible Fun Adaptable Timeless Confident Sophisticated Manly Business casual Cheap Low key Mediocre Average Boring Soft Down to earth Logo Rustic Old brand Historic Great price Global brand Mainstream Steady Goes with everything

Because the two rounds of pretests were conducted in the USA and only Jeep was included in the pretest due to familiarity with Jeep menswear, ten Chinese consumers were interviewed using the same open-ended questions to test whether the associations gathered were consistent with what Chinese consumers had in mind. Final modifications were made, and additional salient associations were added. The word banks for Ford auto and menswear are shown in Tables 4 and 5.

Data Collection

Because there are two brands involved in this study and Ford clothing is relatively new in the Chinese market, different methods were used to collect the data. Specifically, responses about Jeep auto, and maps of brand were collected in the Jeep dealership in Shanghai and responses about Jeep apparel and maps were collected through online interviews. Each participant received a three-section printed or electronic document. The first part was an explanation of the research with a BCM example from the Mayo Clinic (John et al., 2006). The example was translated into Chinese for further data collection.

The second part was a blank page with the brand logo for mapping according to the word bank and images of products of the target brand. Since Ford menswear is a

Table 6 Scale items used to measure brand engagement in self-concept

1. I have a special bond with the brands that I like
2. I consider my favorite brands to be a part of myself
3. I often feel a personal connection between my brands and me
4. Part of me is defined by important brands in my life
5. I feel as if I have a close personal connection with the brands I most prefer
6. I can identify with important brands in my life
7. There are links between the brands that I prefer and how I view myself
8. My favorite brands are an important indication of who I am

new brand, more images collected from its official Web site and online store were included. Participants were encouraged to use their own words and add anything they considered important and necessary that was not included in the provided word bank. This part also included instruction for aggregation to generate a brand concept map. Following the five-step aggregation rules (see Table 1), core brand associations were selected, then first-order associations were selected and linked, non-core brand associations were selected and linked, and in the final step, the number of connecting lines was determined. Associations included were then categorized based on Korchia's association categories.

The third part of the research instrument was a brief survey that includes questions assessing an individual's brand engagement and questions about demographic information. This study also assessed how individual's brand engagement is related to their brand concept maps. To assess individual engagement level, we adopted the scales of brand engagement developed by Sprott et al. (2009). Participants responded on a 7-point Likert scale. Table 6 lists all the scales.

Sample

The study included 60 samples, 15 for Jeep automobile, Ford automobile, Jeep menswear and Ford menswear, respectively. All the respondents were Chinese consumers from major cities (see Table 7).

The age of the sample ranged from 22 to 55, with a mean age of 33.3 years. 80% of the sample was in their 20 s or 30 s. Only a few were over 40 years old. The sample was almost equally distributed across gender with 48.3% of male respondents. Overall, the monthly expenditure on fashion products ranged from 0 to 100 to over 500 yuan; 40% of the sample indicated that they spend more than 500 yuan (~$80) per month on fashion products. Only about 10% of the sample indicated that they spend less than 200 yuan (~$30) per month. 11.7% indicated that their monthly expenditure ranged from 200 to 299 yuan (~$30–50); 20% spent between 300 and 399 yuan (~$50–65), and 16.7% spent between 400 and 499 yuan (~$65–80) per month.

Table 7 Profile of the sample

		Frequency	Percentage (%)
Age (mean = 33.3)	20–29	22	36.7
	30–39	27	45.0
	40–49	8	13.3
	50–59	3	5.0
Gender	Male	29	48.3
	Female	31	51.7
Monthly expenditure on fashion products	¥0–100 (~$0–15)	3	5.0
	¥100–199 (~$15–30)	4	6.7
	¥200–299 (~$30–50)	7	11.7
	¥300–399 (~$50–65)	12	20.0
	¥400–499 (~$65–80)	10	16.7
	over ¥500 (over $80)	24	40.0

Research Findings and Results

Brand Engagement Level

Based on participants' responses, high-engagement and low-engagement groups were identified based on the mean score (4.91) of the overall engagement level. If a respondent's average score was higher than 4.91, he/she was placed in the high-engagement group; if it was lower than 4. 90, he/she was placed in the low-engagement group.

As shown in Table 8, the high-engagement group was composed of 34 respondents (mean 5.75), with 26 respondents (mean 3.82) in the low-engagement group.

Mapping and Results

Four maps were created for the study to evaluate the brand extension success from the Ford automobile to Ford menswear and from the Jeep automobile to Jeep menswear. The maps of the extensions were analyzed by comparing the mention of associations, especially the core associations and interconnections with the parent

Table 8 Engagement level of respondents

Group	Frequency	The level of engagement
High-Engagement	34	5.75
Low-Engagement	26	3.82
Overall	60	4.91

brands. Associations that transferred from the parent brands to the extensions were identified. The associations that were mentioned by the high- and low-engagement groups within each group were also identified, compared, and classified according to Korchia's (1999) categories.

Brand Concept Map for Ford Auto

According to the aggregation rules, the first step identified core brand associations by measuring the frequency of brand associations mentioned and the number of interconnections. Associations mentioned on at least 50% of individual maps were included as core brand associations. Those found on 45–49% of the maps could also be included if the number of interconnections was equal to or higher than the core associations. For the Ford automobile, four associations were selected as core brand associations: Ford Focus, (which is a very popular model of Ford in China), Old brand, American, and Poor gas mileage. The second step was to select first-order associations from the core associations. Frequencies, ratio of first-order mentions and interconnection types of first-order mentions in individual maps, were used as means to determine first-order associations. As shown in Table 9, all of the core associations had a ratio of first-order mentions to total mentions of at least 50% and had more superordinate than subordinate interconnections. Hence, all of them were included as first-order associations in the consensus map (see Table 10).

Table 9 Core associations of Ford auto brand

Associations	Frequency of mention	Number of interconnections
Focus	11	20
Old brand	8	5
American	8	4
Poor gas mileage	10	1

Table 10 First-order associations of Ford auto brand

First-order association	Frequency of first-order	Ratio of first-order mention (%)	Superordinate connections	Subordinate connections
Focus	9	82	2	0
Old brand	6	75	3	0
American	7	88	1	0
Poor gas mileage	6	60	4	0

The third step was to link core brand associations on the map. To determine which links to include, frequencies between specific associations were measured across individual maps. According to the aggregation rules, the inflection point determines the target number of association pairs included. The infection point is where a sharp increase in frequency counts is observed. For instance, for Ford automobile, 21 association pairs had one frequency count, 4 had two counts, and 1 had three. Therefore, the infection point occurred at two and that was considered the target number. Thus, all association pairs that had two or more counts were included. The fourth step was to select non-core association links; based on the rules, all non-core associations that are linked to core associations or directly to the brand, and have a frequency count on or above the target number were included. The core association pairs and non-core association pairs included in the consensus map are shown in Table 11.

The final step was to determine the strength of the lines. When counting frequencies of the core associations and association pairs, the number of lines one, two, or three was also counted. The means of lines of each core association and association pair were then calculated and rounded up or down to the next integer number. Results were presented in Tables 12 and 13.

After the counting and calculating, the consensus map for the Ford automobile was created (Fig. 1). A total of 4 core associations and 4 non-core associations were included, and 13 pairs of links were identified in the 8 associations. As shown in Tables 9 and 10 the association focus, a popular compact model in China, had the highest frequency count and number of interconnections. It was closely associated with Poor gas mileage, affordable, classic and Changan Ford. Changan Ford Automobile Co., Ltd. is a 50–50 joint venture between Ford Motor and Changan Automobile Group established in Chongqing, China, in 2001. It handles all of Ford automobile's business in China, including developing, manufacturing, sales, and service. Another popular model included was Mondeo, a more expensive family car.

Table 11 Frequency of interconnections (core associations in bold)

Interconnections	Frequency
Focus-Poor gas mileage	2
Focus-Changan Ford	2
Focus-Affordable	2
Focus-Classis	3
American-Old brand	2
High quality-Reliable	3
High quality-Durable	3
Durable-Pick up	2
Pick up-Bulky	2
Pick up-Tough	2
Good price-Cost efficient	2
Outdated Style-Mediocre	2
Mondeo-Reputable	3

Table 12 Number of lines of association links included

Interconnections	Frequency	Number of lines	Mean of lines
Focus-Poor gas mileage	2	5	2.5
Focus-Changan Ford	2	5	2.5
Focus-Affordable	2	4	2
Focus-Classic	3	6	2
American-Old brand	2	4	2
High quality-Reliable	3	6	2
High quality-Durable	3	6	2
Durable-Pick up	2	4	2
Pick up-Bulky	2	4	2
Pick up-Tough	2	5	2.5
Good price-Cost Efficient	2	5	2.5
Outdated Style-Mediocre	2	4	2
Mondeo-Reputable	3	3	1

Table 13 Number of lines of associations connected directly to the brand

Associations	Frequency of mention	Number of lines	Mean of lines
Focus	11	29	2.6
Old brand	8	18	2.3
American	8	23	2.9
Poor gas mileage	10	27	2.7
Good price	5	13	2.6
Mondeo	6	10	1.7
High quality	6	12	2
Outdated style	2	2	1

Figure 1 illustrates the final consensus map for Ford automobile. According to the map, participants seemed to have the impression that Ford is associated with Poor gas mileage. Many of them included that association in the maps. They were also aware of the fact that Ford is an iconic, old American brand that is reputable, reliable, safe, historic, and durable. Almost all of the participants who mentioned style and design in their maps indicated negative associations. They used words such as "cheap," "outdated," and "mediocre" to describe the design, interior and exterior of Ford cars. However, non-core associations of "good price" and "high quality" were included as well.

Among the 15 participants, seven were in the high-engagement group and eight were in the low-engagement group based on their engagement scores. Most participants in both groups were in their 30 s, but participants in the high-engagement group tended to spend more on fashion products than did those in the low-engagement group. Although the percentage of participants that had purchased

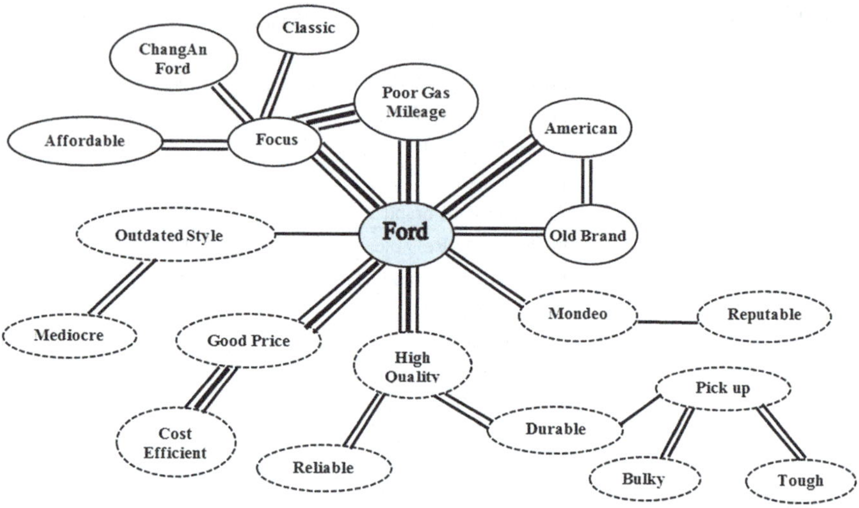

Fig. 1 BCM for Ford auto brand (n = 15)

Ford cars for both groups was below 50%, most of the participants either were considering buying one or had experience driving Ford cars.

Table 14 shows the core associations mentioned by participants in the two engagement groups. "Focus" was the association that had the highest frequency counts and interconnections in both groups. "Poor gas mileage" was also mentioned by both groups. Other than those common core associations, "reliable," "mediocre," and "durable" were identified as core associations in the high-engagement group as well. Participants in the low-engagement group were more aware of and familiar with the different models, as two models, "Mondeo" and "Kuga," were mentioned. They were also more concerned about the price. Other associations important to the low-engagement group were "American" and "Old brand." Table 15 shows the association categories of high- and low-engagement groups for the Ford automobile.

Following the same procedures and steps, BCMs for Jeep auto, Ford menswear, and Jeep menswear were generated.

Brand Concept Map for Jeep Auto

Figure 2 illustrates the final consensus map for Jeep automobile. According to the map, eight core associations were indicated, with "off-road" the most important association. Several of the core associations were connected to the brand through "off-road," such as "off-road" to "Poor gas mileage," "tough" and "four-wheel drive." The model "Wrangler" was closely related to "classic" and "young." "Classic" was also connected to "stylish." Three non-core associations were indicated: "manly," "rugged," and "young."

Table 14 Core associations of different engagement groups

High-engagement group			Low-engagement group		
Associations	Frequency of mention	Number of interconnections	Associations	Frequency of mention	Number of interconnections
Focus	5	12	**Focus**	6	8
Reliable	4	3	**Old brand**	5	3
Mediocre	5	2	**American**	5	0
Durable	5	2	**Poor gas mileage**	5	0
Poor gas mileage	5	1	Mondeo	5	6
			Kuga	4	2
			Cost efficient	4	0
			Good price	4	0

Table 15 Association categories for both engagement groups for Ford auto

	High-engagement group	Low-engagement group
Product-related attributes	Poor gas mileage	Poor gas mileage
Brand personality/lifestyle	Reliable	
Functional benefits	Durable	
The company	Focus	Focus, Mondeo, Kuga
		American, Old brand
Price		Good price, Cost efficient
Attitudes	Mediocre	

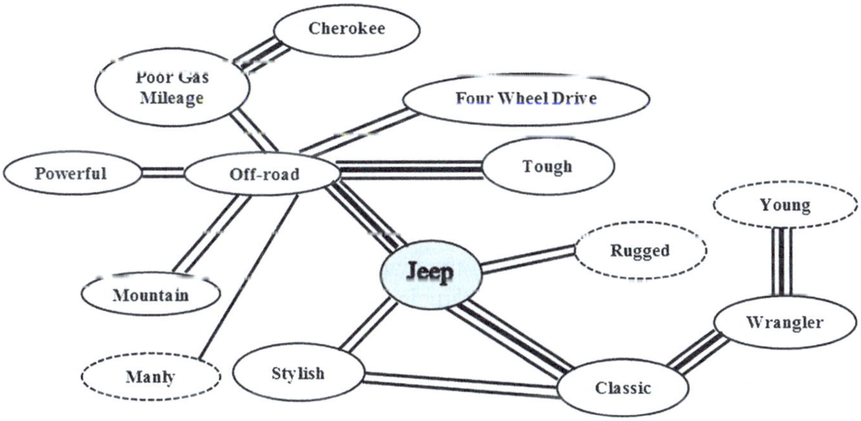

Fig. 2 BCM for Jeep auto brand (n = 15)

Table 16 Association categories for both engagement groups for Jeep auto brand

Brand associations	High-engagement group	Low-engagement group
Product-related attributes	Poor gas mileage	Poor gas mileage
	Four-wheel drive	4-wheel drive
Functional benefits	High quality	
Experiential benefits	Drives well	
The company		Cherokee, Wrangler
Symbolic benefits	Classic, Stylish	Classic
Brand personality/lifestyle		Tough, Manly
Usage imagery	Off-road	Off-road

Based on the engagement scores, five participants were in the high-engagement group and ten participants were in the low-engagement group. Participants in the high-engagement group tended to spend more on fashion products, and 80% had purchased Jeep vehicles. Participants in the low-engagement tended to distribute more evenly on different levels of monthly expenditure on fashion products. Around 40% of the participants in the low-engagement group had purchased Jeep vehicles.

The core associations mentioned by both groups were "off-road," "classic," "Poor gas mileage," and "four-wheel drive." In addition to the four common core associations, the high-engagement group also mentioned "high quality," "drives well," and "stylish." The low-engagement group made more core associations than the high-engagement group. Other important associations mentioned were "tough," "rugged," "manly," "Wrangler," and "Cherokee." The core associations for both groups were categorized (Table 16).

Brand Concept Map for Jeep Menswear

Figure 3 illustrates the final consensus map for Jeep menswear. In the map, five core associations were indicated ("manly," "outdoorsy," "durable," "Jeep car," and "tough"). The two non-core associations indicated were "American" and "high quality." According to the map, "Jeep car" seemed to be the most important core association, as it was connected to many other associations, such as "sporty," "durable," and "tough." "Jeep car" also had the most interconnections, number of lines, and the highest frequency of first-order mention.

Twelve of the participants were in the high-engagement group and three in the low-engagement group. Most of the participants were in their 20 and 30 s. In the high-engagement group, nine were female and three were male. In the low-engagement group, two were female and one was male. All of the participants at least had visited and shopped in Jeep menswear retail stores or online. The male participants had at least tried on some pieces. Almost half of the participants in the high-engagement group had purchased Jeep menswear clothing. Participants in the low-engagement group had either purchased Jeep vehicle or clothing.

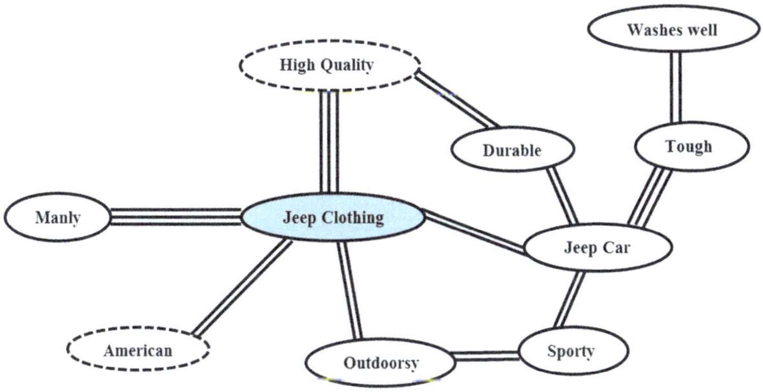

Fig. 3 BCM for Jeep menswear brand (n = 15)

Four associations were mentioned by both groups: "durable," "manly," "outdoorsy," and "tough." These four associations were also identified as core associations in the final consensus map. Among those, "tough" and "manly" were cited as core associations in both the Jeep automobile and Jeep menswear maps. The core association in the high-engagement group that had the highest frequency count and the number of interconnections was "Jeep Car," whereas in the low-engagement group, participants rated other associations such as "everyday wear," "sporty," and "American" higher. Table 17 illustrates the association categories of both engagement groups.

Brand Concept Map for Ford Menswear

Figure 4 shows the final consensus map for Ford menswear. According to the map, Ford menswear was associated closely with "Ford car" and "American." These two core associations were linked to other positive associations such as "reputable," "global brand," and "high quality." Other associations addressed the pricing, design, and cut of the line.

Ten participants were in the high-engagement group, and five were in the low-engagement group. Most of the participants in both engagement groups were in their 20 s. Half of the participants in the high-engagement group were male and the

Table 17 Association categories for both engagement groups for Jeep menswear

Brand Associations	High-engagement group	Low-engagement group
Functional benefits	Durable	Durable, High quality
The company	Jeep Car	American
Brand personality/lifestyle	Tough, Manly	Tough, Manly
Usage imagery	Outdoorsy	Outdoorsy, Sporty, Everyday Wear

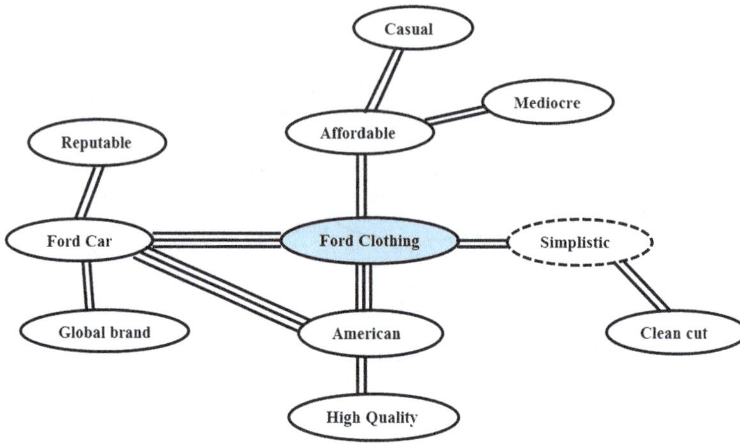

Fig. 4 BCM for Ford menswear brand (n = 15)

rest were female. In the low-engagement group, only one was male and four were female. Monthly expenditure on fashion products in the high-engagement group tended to evenly distribute from approximately 0–100 yuan (~$0–15) to over 500 yuan (~$80). Participants in the low-engagement group spent at least 200 yuan (~$30) monthly on fashion products.

None of the participants owned a Ford vehicle. Also, as the clothing line was newly launched in selective cities, none of the participants had a chance to make any purchase yet. However, all of the participants were familiar with the Ford brand, and all had experience driving or riding in a Ford car.

Both groups cited "mediocre" and "American." The low-engagement group also mentioned "Old brand." "Ford car" had the greatest number of interconnections and high-frequency counts in the high-engagement group. Table 18 shows association categories for both engagement groups.

Table 18 Association categories for both engagement groups for Ford menswear

Brand associations	High-engagement group	Low-engagement group
Price	Affordable	
The company	Ford car, American	American, Old brand
Attitudes	Mediocre	Mediocre
Brand personality/lifestyle	Simplistic	

Brand Extension Evaluation and Mapping Comparison

According to the maps and Korshia's (1999) association categories, participants connected the Jeep automobile to Jeep menswear easily. Many unique and positive associations with the Jeep automobile transferred to the menswear extension successfully and smoothly. In particular, associations of categories such as brand personality/lifestyle, user imagery, the company and functional benefits transferred completely from the parent brand to the extension.

Similar to the case with Jeep, associations with the Ford motor company and country of origin transferred well from the parent brand to the extension. Other association categories that transferred successfully were the price and attitudes. Although as an iconic American brand, Ford was considered "reputable," "high quality," and "durable," some consumers used "mediocre" to describe both the Ford automobile and its menswear. Compared to the case with Jeep, as a parent brand, Ford lacked a positive and unique brand personality. Hence, the extension lacked identity as well. Product-related association and functional benefits did not transfer to the extension at all. The affordable price of the menswear line was associated directly with the negative attitude of "mediocre."

Discussion and Implications

This study demonstrated that BCM is an effective and powerful tool in the evaluation of brand extension, as it offers a clear picture of consumers' association networks for both parent brands and extensions. The visual format enables researchers and practitioners to trace which and how associations transfer from parent brands to extensions and make comparisons easily. Such market research can help manufactures who are searching global brands for licensing to select correct brands. For those licensors, such type of research can help them react quickly and adjust strategically according to the mapping results.

The generated BCMs revealed that, in the case that fit between automobile and fashion is poor, strong associations, both positive and negative, still transferred well from the parent brands to their extensions. Thus, it is clear that associations with the parent brand have a great influence on the extension, even between distant product categories. For instance, the associations of "good price," "affordable," and "mediocre" were identified in the Ford automobile map and then as core associations in the Ford menswear map.

The results showed that brand personality could be a key factor in the success of a fashion brand extension. For instance, associations with the Jeep automobile, such as "tough" and "manly," transferred successfully to the menswear line and became the core associations. Compared to Jeep, the new Ford menswear line lacked the strong brand personality-related associations that need to be leveraged in order to create a strong brand positioning for Ford menswear. However, as shown in the

Ford menswear map, "simplistic" and "clean cut" offer a hint to marketers with respect to how they might change consumers' overall perceptions of the brand Ford as being "mediocre."

As shown in the maps, in both the automobile and fashion categories, the country of origin effect played an important role in brand evaluation and the success of the extension; "American" was mentioned in three of the four maps as a core association, especially in the extensions, although neither was made or designed in the USA. "American" was linked to positive associations, such as "high quality", so firms should capitalize on the country of origin effect when promoting their extensions. This effect may be particularly powerful in emerging markets.

In addition, the results showed that the level of brand engagement had a positive association with consumer's expenditure on fashion products. Firms should locate the group of consumers in the high-engagement level, analyze their association maps, and tailor their products to those consumers' needs. As the results suggest, firms that are considering extending their business into the fashion category should examine the brand association networks of the parent brand thoroughly and determine on which associations they should capitalize. The results revealed that the success of Jeep was not a coincidence, as so many positive key associations transferred from the parent brand to the extension. In contrast, Ford needs to create more positive associations regarding brand personality and functional benefits to change consumers' negative attitudes.

Another important implication for manufactures from emerging markets to leverage a global brand is that the country of origin effect should always be assessed. Consumers in emerging markets are more aware of brand names that originate in developed market and are more likely to generate positive associations accordingly. Both the case of Jeep and Ford supported this assertion. Particularly, in the Ford menswear map, the association "American" was linked directly to "high quality." Therefore, we suggest that a new extension with a relatively weak brand image should take advantage of the country of origin effect when first entering the market. Meanwhile, when making licensing decision, apparel manufactures should evaluate a brand's personality and identity and to what degree the brand' apparel extension can express consumers' self, identity, and personality.

Limitations and Future Studies

The study has several limitations, primarily that it cannot be generalized well due to the sample size (15 for each map), and the fact that convenient/snowball sampling was used. Thus, the samples do not represent Chinese consumers in Shanghai, Beijing, Tianjin, and Shenzhen adequately. In particular, in the case of Ford menswear, the sample may be biased because Ford menswear has no retail stores in those cities. The results also cannot be generalized to the fashion industry, as only two brands of menswear were selected and they may not represent the entire category. Wording problems also occurred, as participants were encouraged strongly to use their own words. The various ways in which participants expressed the ideas and thoughts in their minds, added difficulty in the aggregation state. This generated a large number of association links with low-frequency counts. Accordingly, these maps might not be as representative as desired. Further, research should continue to apply BCM to evaluate extensions in other industries. Cross-cultural analysis would also be a good direction for comparison. The role that brand engagement plays in brand extension should be studied further as well.

References

Aaker, D. A. (1991). *Managing brand equity: Capitalizing on the value of a brand name*. New York/Toronto/New York: Free Press/Maxwell Macmillan Canada/Maxwell Macmillan International.

Aaker, D. A., & Keller, K. L. (1990). Consumer evaluations of brand extensions. *Journal of Marketing, 54*(1), 27–41.

Anderson, J. R. (1983). *The architecture of cognition* (p. 1983). Cambridge, Mass.: Harvard University Press.

Anderson, J. R., & Bower, G. H. (1980). *Human associative memory: A brief edition*. Hillsdale, N. J.: L. Erlbaum Associates.

Bottomley, P. A., & Holden, S. J. S. (2001). Do we really know how consumers evaluate brand extensions? Empirical generalizations based on secondary analysis of eight studies. *Journal of Marketing Research, 38*(4), 494–500.

Boush, D. M., & Loken, B. (1991). A process-tracing study of brand extension evaluation. *Journal of Marketing Research, 28*(1), 16–28. https://doi.org/10.2307/3172723.

Bristol, T. (1996). Consumers' beliefs resulting from conceptual combinations: Conjunctive inferences about brand extensions. *Psychology & Marketing, 13*(6), 571–589.

Carroll, A. (2009). Brand communications in fashion categories using celebrity endorsement. *Journal of Brand Management, 17*(2), 146–158. https://doi.org/10.1057/bm.2008.42.

Dew, L., & Kwon, W.-S. (2010). Exploration of apparel brand knowledge: Brand awareness, brand association, and brand category structure. *Clothing & Textiles Research Journal, 28*(1), 3–18. https://doi.org/10.1177/0887302X09341877.

Escalas, J. E. (2004). Narrative processing: Building consumer connections to brands. *Journal of Consumer Psychology, 14*(1/2), 168–180.

Escalas, J. E., & Bettman, J. R. (2005). Self-construal, reference groups, and brand meaning. *Journal of Consumer Research, 32*(3), 378–389.

Fiske, S. T. (1982). *Schema triggered affect: Applications to social perception*. Paper presented at the Affect and cognition: 17th Annual Carnegie Mellon symposium on cognition.

Fiske, S. T., & Pavelchak, M. A. (1986). Category-based versus piecemeal based affective responses: Developments in schema-triggered affect. In R. M. Sorrentino, E. T. Higgins, R. M. Sorrentino, & E. T. Higgins (Eds.), *Handbook of motivation and cognition: Foundations of social behavior* (pp. 167–203). New York, NY, US: Guilford Press.

French, A., & Smith, G. (2013). Measuring brand association strength: A consumer based brand equity approach. *European Journal of Marketing, 47*(8), 1356–1367. https://doi.org/10.1108/03090561311324363.

Hem, L. E., De Chernatony, L., & Iversen, N. M. (2003). Factors influencing successful brand extensions. *Journal of Marketing Management, 19*(7–8), 781–806.

Henderson, G. R., Iacobucci, D., & Calder, B. J. (1998). Brand diagnostics: Mapping branding effects using consumer associative networks. *European Journal of Operational Research, 111* (2), 306–327.

Jayachandran, S., Kaufman, P., Kumar, V., & Hewett, K. (2013). Brand licensing: What drives royalty rates? *Journal of Marketing, 77*(5), 108–122.

John, D. R., Loken, B., Kim, K., & Monga, A. B. (2006). Brand concept maps: A methodology for identifying brand association networks. *Journal of Marketing Research, 43*(4), 549–563.

Keller, K. L. (1993). Conceptualizing, measuring, and managing customer-based brand equity. *Journal of Marketing, 57*(1), 1–22.

Keller, K. L. (2008). *Strategic brand management: Building, measuring, and managing brand equity*. Harlow: Pearson.

Klink, R. R., & Smith, D. C. (2001). Threats to the external validity of brand extension research. *Journal of Marketing Research, 38*(3), 326–335. https://doi.org/10.2307/1558527.

Korchia, M. (1999). A new typology of brand image. *ACR European Advances, 4,* 147–154.

Levy, S. J. (1981). Interpreting consumer mythology: A structural approach to consumer behavior. *Journal of Marketing, 45*(3), 49–61.

Levy, S. J. (1985). Dreams, fairy tales, animals, and cars. *Psychology & Marketing, 2*(2), 67–81.

Martinez, A. (2011). *Consumer evaluations of brand extensions: Can online brands be extended to offline markets?* (MS), The Heriot-Watt University, Edinburgh, United Kingdom.

Morein, J. A. (1975). Shift from brand to product line marketing. *Harvard Business Review, 53*(5), 56–64.

Priluck, R., & Till, B. D. (2010). Comparing a customer-based brand equity scale with the implicit association test in examining consumer responses to brands. *Journal of Brand Management, 17* (6), 413–428. https://doi.org/10.1057/bm.2009.32.

Romeo, J. B. (1991). The effects of negative information on the evaluations of brand extensions and the family brand. *Advances in Consumer Research, 18*(1), 399–406.

Shin, S. (2011). *Application of brand concept maps to GAP Inc.* (MS), University of Georgia.

Shocker, A. D., Srivastava, R. K., & Ruekert, R. W. (1994). Challenges and opportunities facing brand management: An introduction to the special issue. *Journal of Marketing Research, 31*(2), 149–158.

Sprott, D., Czellar, S., & Spangenberg, E. (2009). The importance of a general measure of brand engagement on market behavior: Development and validation of a scale. *Journal of Marketing Research, 46*(1), 92–104.

Susan, A., & Richard, E. (1998). Fashion involvement, self-monitoring and the meaning of brands. *Journal of Product & Brand Management, 7*(2), 109–123. https://doi.org/10.1108/10610429810216874.

Völckner, F., & Sattler, H. (2006). Drivers of brand extension success. *Journal of Marketing, 70* (2), 18–34.

Walsh, P., Rhenwrick, I., Williams, A., & Waldburger, A. (2014). Team brand extension or licensed product? Examining consumer awareness of two distinct brand strategies. *SBM Sport, Business and Management: An International Journal, 4*(2), 96–108.

Wyer, R. S., Budesheim, T. L., Lambert, A. J., & Swan, S. (1994). Person memory and judgment: Pragmatic influences on impressions formed in a social context. *Journal of Personality and Social Psychology, 66*(2), 254–267. https://doi.org/10.1037/0022-3514.66.2.254.

Zhuang, G., Wang, X., Zhou, L., & Zhou, N. (2008). Asymmetric effects of brand origin confusion evidence from the emerging market of China. *International Marketing Review, 25*(4), 441–457.

Part V
Select Fashion Market Segments in China

Chinese Male Consumers' High-End Shirt Consumption: A Perspective of Fabric Attributes

Yanwen Ruan, Yingjiao Xu, Jun Li and Xiaogang Liu

Abstract With rising disposable incomes and male consumers' increasing interest in personal grooming products, the Chinese menswear market is growing tremendously. Shirt is one of the most popular menswear items chosen by Chinese male consumers. While the shirt market is growing and many Chinese fabric manufacturers possess great capacity in producing high-end fabrics and supply to globally known upscale shirt brands, no fabric brand is in existence in China for high-end shirts. Appreciating the additional value from the branded ingredient (fabric), especially for upscale products (shirts), it might be to the great benefits of the current large fabric enterprises to brand their fabric products. To meet the increasing demand for high-end men's shirts in the Chinese market and to help Chinese textile industry establishing high-end shirt brands, it is important to understand Chinese male consumers' shirt consumption behavior for high-end shirts. Focusing on consumers' preference of fabric attributes, this study conducted a two-step cluster analysis with a survey of 170 Chinese male consumers. As a result, three distinct clusters in the Chinese men's shirts market were identified: premium consumers, pragmatic consumers, and economic consumers. Results of this study will not only provide implications to the practitioners in the industry in their efforts to better meet the Chinese male consumers' needs for high-end shirts, but also contribute to the literature on male consumers' shirt consumption behaviors.

Keywords Male consumers · High-end shirts · Fabric features · Cluster

---

Y. Ruan (✉) · J. Li · X. Liu
Donghua University, Shanghai, China
e-mail: ruanyanwen.elsa@hotmail.com

Y. Xu
North Carolina State University, Raleigh, USA

Y. Xu et al. (eds.), *Chinese Consumers and the Fashion Market*, Springer Series in Fashion Business, https://doi.org/10.1007/978-981-10-8429-4_8

Introduction

In the global apparel and footwear market, Asia Pacific, North America, and Western Europe represent the three largest markets (Euro Monitor, 2017a; Prnewswire, 2016). In 2016, the Asia Pacific menswear market had exceeded the other two markets. Meanwhile, from 2002 to 2016, Western Europe had showed a declined tendency in menswear market (Euro Monitor, 2017b; Mintel, 2017). Representing the largest part of the menswear market in Asia Pacific, China's menswear industry grew much faster than the womenswear industry. In the menswear market, the shirt segment would continue to grow in the years to come (Bizvibe, 2015; China Luxury Advisors, 2015; Edited, 2016; Euro Monitor, 2017d). As a commonly utilized item, shirts play a very important role in male consumers' daily life and are the second largest category among all made-to-measure categories (China Business Industry Research, 2017a).

In the category of menswear shirts, high-end shirts such as ready-to-wear or made-to-measure formal shirts and dress shirts traditionally take the main part in male consumers' wardrobe (Storey, 2008). Although there is not a common definition for high-end shirts, this category can be identified from three aspects. From the view of price level, a high-end shirt is usually priced over USD150 in the USA (Gentleman's Gazette, 2013). From the view of dress occasion, high-end shirts are usually dressed in formal occasions (Storey, 2008), such as wedding, evening ball, or job interviews. From the view of resource and craftsmanship, high-end shirts usually made by good quality fabric, long staple cotton, such as GIZA cotton from Egypt. Some high-end shirts have logo of the high-end fabric brand on their shirts to display their luxury taste (Gentleman's Gazette, 2010). Not only the fabric, but also the high-quality accessories, the exact fit tailor, and the excellent craftsmanship are all important to a high-end shirt (Guokr, 2014).

A national survey indicated that 61% of US male Internet users aged over 18 have purchased dress shirts in 2016, just followed common casual wear T-shirts, jeans, pants, shorts, and polo shirts (Mintel, 2016). In China, the formal dress shirts and the made-to-measure shirts markets will continue to develop (Pinpai China, 2017).

While the demand for men's high-end shirts is strong, no Chinese high-end shirt fabric brand is in existence. In comparison, the manufacturing of high-end shirt fabrics in China is very well developed, but their production is mainly to supply to the global fashion brands. The well-known high-end shirt fabric brands around the world are mainly owned by European companies, including Albini Group, Canclini, Testa, Monti, Alumo, Soktas, and Getzner. To meet the demand for high-end men's shirts in the Chinese market, it is important for the industry to understand Chinese male consumers' shirt consumption behaviors, particularly their preference of fabric attributes.

This current study is designed to provide an insight of Chinese male consumers' high-end shirt consumption. Specifically, it aims to investigate Chinese male consumers' preference of fabric attributes for high-end shirts. A two-step cluster analysis was conducted to further segment the Chinese high-end shirt market. Results of this study not only provide implication to the practitioners in the industry in their efforts to better meet the Chinese male consumers' needs for high-end shirts, but also contribute to the literature on male consumers' shirt consumption behaviors.

In the following sections, an overview of the global and Chinese menswear market was first provided and followed by an introduction and analysis of the global high-end men's shirt fabric brands. The literature review section ends with a review of male consumers' shirt consumption. Following the literature review, the research methodology is explained in detail. Finally, the findings of this study are presented in the results section, and conclusions and implications are provided at the end of the paper.

Literature Review

Global Menswear Market

The global menswear market sized around USD414.2 billion in 2016, and the volume of the market will reach around USD450 billion in 2021 (Euro Monitor, 2017a). The Asia Pacific represents the largest menswear market in the world, followed by North American and Western Europe. As male consumers continue to place a greater emphasis on their appearance, menswear is expected to outpace womenswear in terms of growth (Business of Fashion, 2014; Euro Monitor, 2017b; Fashion United, 2016; Rinallo, 2007).

North America was the second largest menswear market in the world, and the USA was the major market in this region with USD87.5 billion sales volume in 2016. Shirts and jumpers were the second largest category in the US menswear market in 2016 (Euro Monitor, 2017c). The US menswear industry was expected to slightly increase in the future (Euro Monitor, 2017c), so was the men's shirt category, which enjoyed a 12.5% volume growth over the period of 2011–2016 (Euro Monitor, 2017c).

Germany, UK, France, and Italy were the leading menswear markets in Western Europe, accounting for 63% of the region's sales of USD90.1 billion in 2016. However, the general menswear in Western Europe has shown a sign of decline (Euro Monitor, 2017b). On the contrast, the shirts market enjoyed an absolute growth of 575 million dollars (Euro Monitor, 2017b). Turkey, Sweden, and Norway were the three largest shirt markets in Western Europe, with Italy, Germany, and France having the smallest shirt markets in this region (Euro Monitor, 2017b).

Table 1 Main menswear industrial clusters in China

Industrial cluster	Number of companies	Production (billion units)	Total value (CNY billion)	Total value (USD billion)	Representative menswear brands
Jiangsu	Over 3000	4.77	77	11.85	Bosideng
Zhejiang	Over 5449	4.01	62	9.54	Saint Angelo, Youngor
Guangdong	Over 1400	6.63	9	1.38	Jeanwest
Fujian	Over 3500	4.09	47.3	7.27	Lilanz, Septwolves

Data source China Business Industry Research (2017a, 2017b)

Menswear Market in China

With CNY51.1 billion (equivalent of USD7.69 billion) of sales in 2016, China's menswear market is the most dominant in Asia Pacific, which represents the largest menswear market in the world (China Luxury Advisors, 2015). The market was expected to continue to grow in the years to come (Euro Monitor, 2017d), accredited to the growth of net income per capita in China. However, compared to other developed regions, the per capita spending on menswear is still low in China (EFU, 2017).

In contrast, the manufacturing sector in China's menswear industry is quite well developed. There is a heavy concentration of the menswear production in the east coast of the country, including the provinces of Jiangsu, Zhejiang, Fujian, and Guangdong. Table 1 summarizes the four main menswear industrial clusters in China. Among these clusters, the Zhejiang cluster leads the shirt and suits category, with Youngor being the leading brand.

China's men's shirts industry enjoyed a total value growth of 29.2% from 2011 to 2016, following tops, jackets and coats, and jumpers (Euro Monitor, 2017d). In the next five years, the shirts industry will continue to grow while the pace will slow down at 4.1% (Euro Monitor, 2017d; Prnewswire, 2016). A similar growth pattern was observed from the sales of the five main Chinese menswear brands over the period of 2014–2016 (Fig. 1). In general, among the five brands, the income from shirts has an upward trend, but the increase slope gets flatter.

Global High-End Shirt Fabric Brands

A typical model for running a shirt brand is to design their own shirt styles, source the fabric from a fabric supplier, contract out the production to a garment manufacturer, and market the products to the consumers. Therefore, many globally known shirt companies do not possess fabric production capacity, instead they work

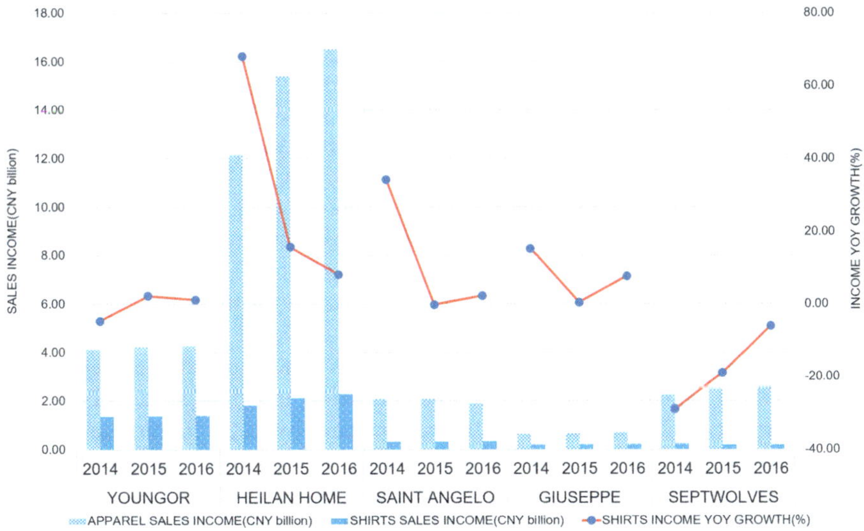

Fig. 1 Sales value of shirts from the five main menswear enterprises in China data. *Source* 2014–2016 annual reports of Youngor, Heilan Home, Saint Angelo, Giuseppe, and Septwolves

with vendors to supply them with needed fabrics. The suppliers could be pure fabric manufacturers not providing design inputs, or fabric brands who design and market (with or without production capacity). China supplies a great amount of fabric to the world; however, very few fabric brands exist in China and there is no high-end shirt fabric brand.

In comparison, there are some globally known brands in the high-end shirt fabric market, with most of them originated in the western countries. Some of them come from Italy, including Albini group, Canclini, Testa, and Monti. Others include Alumo from Switzerland, Soktas from Turkey, and Getzner from Austria. These brands possess unique competitive advantages in the shirt fabric market, including heritage advantage, raw materials advantage, and innovation advantage.

Most of the globally known high-end fabric shirt brands were established a century ago. The built-up expertise on fabric design, production, and marketing provides a great foundation to compete in the high-end shirt fabric market. Raw material is the most basic and important feature of an upscale fabric. All these upscale fabric brands, which are Albini, Canclini, Testa, Monti, Alumo, Soktas, and Getzner, use some types of high quality and scarce raw materials, including GIZA 45 and GIZA 87 cotton from Egypt, West Indies Sea Island cotton, MEANDER71, Pima cotton, and European Norman Flax.

The competitive advantages of the global upscale fabric brands take in the form of innovations in pattern, color, function, and texture. As to the pattern innovation advantage, most of those global brands invest on the development of over thousands of different patterns to meet the needs of consumers (Albini, 2017; Alumo, 2017; Canclini, 2017; Getzner, 2017; Monti, 2017; Soktas, 2017; Testa, 2017;). For example, Getzner develops over 5000 different patterns every year, and Canclini provides over 3500 new pattern designs in every six months in its innovation laboratory (Canclini, 2017; Getzner, 2017). The color innovation advantage refers to the ability to create great amount of color every year. Most of the upscale fabric brands not only pay close attention to the color trend, but also develop their own unique colors to meet their customers' needs (Monti, 2017). The functionality innovation advantage usually refers to the finishing effect of the fabric. In addition to the function of permanent press, new functions include self-decontamination, temperature control. (Albini, 2017; Alumo, 2017; Canclini, 2017; Getzner, 2017; Soktas, 2017). The texture advantage refers to the use of high-count fabric. The highest count of fabric is 400/2, which is currently offered by Getzner. With the high count comes with new fabric features, such as hand and feel (Getzner, 2017).

While there are some large-scale textile enterprises in China, such as Lutai and Lianfa, and most of them have great innovation capacity, according to the analysis of eight major Chinese cotton textile corporates (China Textile Planning Research Association, 2014), there is no upscale shirt fabric brand in China. With the strong demand for high-end shirts for male consumers, there is also a market for high-end fabric brands. These large enterprises have the great potential to develop their own fabric brands in terms of access to high-quality raw materials (enabled by the great financial resources available), innovation ability, and historic heritage. A critical step in developing a successful brand is to understand consumers' needs as well as their preferred product attributes.

Male Consumers and Shirts

In the West, the commercial representations of masculinity have shifted from hard work, thrift, and production to personal pleasure, stylistic expression, and commodity consumption (Osgerby, 2003; Shaw & Tan, 2014). In menswear, shirts are a popular item for hundred years. Among all kinds of shirts, the white formal shirts used to symbolize power, responsibility, success, wealth, and reliability. Although dress shirts are not as popular as in the past, they are still the must-have items close to men's personal life and professional life today and still a symbol of masculinity.

According to the analysis by Mintel, men want more flexibility in work hours, especially young male adults (Mintel, 2017), and not many companies strictly require formal wears in workplaces now (Euro Monitor, 2017b). Therefore, casual shirts and jeans are accepted by more male consumers in workplaces. Shirts other than in white color are more accepted by male consumers also.

In terms of materials, according to the qualitative research on the twenty-first century fashion tailors by Ross (2007), most fashion tailors claimed that made-to-measure customers still prefer suits made in natural fibers. There is a demand for lighter fabrics with more creative colors and patterns, and with natural blended fibers (Ross, 2007). Although Ross' research was based on suits, the preference for upscale shirts is expected to be similar. Additionally, composition and pattern are another two features that should be considered in future research.

Male consumers in China showed a stronger interest in buying personal grooming products and had been spending more time shopping for themselves, especially shopping for apparel products (Bakewell, Mitchell, & Rothwell, 2006). Shirt is the most popular item chosen by Chinese male consumers for both casual and formal occasions (China Business Industry Research, 2017a, 2017b). Among all apparel categories, the consumers of shirts are the youngest, with 65% under 24 years of age (China Business Industry Research, 2017a, 2017b). Young men of age under 24 pay more attention to apparel products with both formal style and pop elements, and shirts and suits are their top choices (China Business Industry Research, 2017a, 2017b; Euro Monitor, 2017a, 2017b, 2017c, 2017d).

Men were found putting more emphasis on convenience when shopping and showed a higher level of loyalty to brands than their female counterparts (Hart, Farrel, Stachow, Reed & Cadogan, 2007). Most of male consumers regarded functionality and practicability as the most important factors for their apparel purchases (Frith & Gleeson, 2004). The menswear industry puts ongoing emphasis on developing functional shirts to meet male consumers' needs for function as well as convenience. For example, durable press (DP) shirts account for 55.5% of whole volume of shirts at Youngor (2014, 2015, 2016). Similar observations were made from other menswear companies in China.

With the growth of personal income and shirt consumers getting younger, the functionality of shirts seemed no longer to be the only feature that menswear companies should improve (EFU, 2017). With more male consumers paying attention to unique custom service, branding, material, design, and cut, customization need is on the rise for menswear (China Business Industry Research, 2017a). Among all made-to-measure categories, shirt is the second most popular item, only after the most popular one—suit (China Business Industry Research, 2017a). With the rise of the need to have unique styles on top of functionality, it is important for the Chinese menswear industry to examine Chinese male consumers' shopping decision for menswear, particular for shirts (EFU, 2017).

Methodology

Data for this study were collected via an online survey. The survey was distributed through a survey distribution agent in China named SOJUMP. A convenience sampling method was used to recruit male consumers from the large national consumer panel that SOJUPM maintains. Consumers must be 18 years of age or older to participate. The online survey took place in June, 2016, and lasted for about four weeks. The data collection process was terminated when the responses seemed sporadic. Extending the survey duration would not increase the response rate greatly. A total of 170 complete questionnaires were received at the end of the data collection period.

Survey Instrument

The questionnaire included three sections. The first section was to measure consumers' general shopping behaviors toward high-end shirts, including purchase frequency, purchase intention, preferred retail channels, and consumption experience. The second section was to measure consumers' preferences of fabric attributes for high-end shirts. Particularly, as studied in previous research, the following fabric attributes were deemed important to shirts and investigated in this study: color (Elliot & Niesta, 2008; Niesta, Elliot & Feltman, 2010; Nippon Color & Design Research Institute Inc., 2017), pattern (Ross, 2007), function (Frith & Gleeson, 2004), and composition/material (Ross, 2007). Table 2 lists the options for each fabric attribute. The last section of the questionnaire measured demographics, including age, income, education, overseas experience, and occupation. The oversea experience was also included in the survey as global brands occupied a large share in the Chinese men's shirts market.

Data Analysis

Cluster analysis (CA) was an exploratory data analysis tool for organizing observed data into meaningful taxonomies, groups, or clusters (Wrenn, Steven & Loudon, 2002). Using cluster analysis, a customer 'type' could represent a homogeneous market segment (Banerjee & Agarwal, 2013). Identifying their particular needs in that market allows products to be designed with greater precision and direct appeal within the segment (Wrenn et al., 2002). This study used the two-step cluster

Table 2 High-end shirt fabric attributes

Fabric attributes	Attribute options
Color	1. Grayscale (black, white, gray) 2. Classic dark (dark blue, dark brown) 3. Classic bright (blue, beige) 4. Unique color (purple, red, green)
Pattern	1. No pattern 2. Stripe and check 3. No stripe or check (flower, geometric, etc.) 4. Obscure pattern (flax-like, denim, plain-colored jacquard)
Functionality	1. Do not care about function 2. Protection 3. Comfort 4. Easy care
Composition/material	1. Do not care about composition 2. Natural fiber or blends 3. Blends including man-made fibers

analysis method in segmenting the male participants based on their preference on the four shirt fabric attributes, namely color, pattern, functionality, and composition.

To validate the cluster solution, three main measures are usually checked for the analysis (Rundle, Kubacki & Tkaczynski, 2015). First, the silhouette measure of cohesion and separation is required to be above the level of 0.0, suggesting valid within-cluster distance and between-cluster distance (Rundle et al., 2015). Second, chi-square tests for categorical variables or t test/one-way ANOVA analyses for continuous variables are conducted to test the validity of the solution (Rundle et al., 2015; Sarstedt & Mooi, 2014). Third, the final cluster solution must be similar when halved (Rundle et al., 2015). For the third validation measure, researchers also suggest using the ratio of sizes of the clusters being less than three as an indicator of good cluster quality (Gaskin, 2012).

When the final cluster solution was generated, further post hoc chi-square analyses were conducted to identify differences among the three clusters in terms of other variables, including demographics and their consumption behavior for high-end shirts. These further variables together with their preferred fabric features were used to develop the profiles for each cluster.

Results

Sample Profile

Descriptive analyses were conducted to compile the sample profile. As presented in Table 3, almost half of the respondents were in the age range of 25–35, with the other half distributed among the remaining age groups. As to income, about 20% of the respondents had a monthly income of more than USD3,600, with about 40% having monthly income between USD1,300 and 3,600 and the other 40% below USD1,300. More than half of the respondents were professionals. The sample in general was very well educated, with more than 85% of the respondents having at least some college education. More than three quarters of the sample had living abroad experiences. In terms of high-end shirt consumption experience, almost 60% of the respondents had purchased shirts at a price higher than CNY500 (USD78) and were willing to purchase shirts priced higher than CNY1,000 (USD150). In terms of preferred shopping channels for high-end shirts, about three quarters of the respondents indicated specialty stores or brand shops in shopping centers. About 12% indicated preference to use private bespoke for their high-end shirt needs. The remaining 14% mentioned online shopping. The majority of the respondents (70%) purchased high-end shirts less often than quarterly, with the remaining respondents had a purchase frequency of at least quarterly or monthly.

Table 3 Sample profile

Variable	Scale	Count	Frequency (%)
Age	18–24	11	6.47
	25–34	74	43.53
	35–44	28	16.47
	45–54	34	20.00
	Over 55	23	13.53
Income monthly (USD)	Less than 1,300	72	42.35
	1,300–2,300	49	28.82
	2,300–2,700	9	5.29
	2,700–3,600	7	4.12
	Over 3,600	33	19.41
Career	Student	12	7.06
	Teacher/faculty	12	7.06
	Industrial professionals	93	54.71
	Freelancer	38	22.35
	State employee	15	8.82
Education	High school diploma or less	23	13.53
	College degree	100	58.82
	Advanced degree	47	27.65
Oversea background	No	130	76.47
	Yes	40	23.53
High-end shirt consumption behaviors			
Purchased shirts at CNY500+ (USD78+)	No	69	40.59
	Yes	101	59.41
Purchase intention of shirts priced CNY1,000+ (USD150+)	No	69	40.59
	Yes	101	59.41
Preferred retail channel	Boutique or brand store	127	74.71
	Private bespoke	21	12.35
	Online	22	12.94
Purchase frequency	Monthly	19	11.18
	Quarterly	32	18.82
	Less often than quarterly	119	70.00

Cluster Analysis Results

Clusters Identified

A two-step cluster analysis based on the four fabric features produced a three-cluster solution (Table 4). There were 42 (24.7%) consumers in the first cluster, with 45 (26.5%) consumers in the second cluster and 83 (48.8%) in the third cluster. The solution had a silhouette measure of cohesion and separation value of greater than 0.2, indicating a satisfactory cluster quality. A chi-square test of the four fabric features suggested significant differences among the three clusters. Lastly, the smallest cluster and the largest one had a size of 42 and 83, respectively, yielding a ratio of cluster sizes of 1.98 (less than three) for the three cluster solution. Thus, all three validation measures suggest a good quality three-cluster solution.

Table 4 Three clusters based on fabric feature preferences

Fabric preference features	Cluster 1 ($n = 42$, 24.7%)	Cluster 2 ($n = 45$, 26.5%)	Cluster 3 ($n = 83$, 48.8%)
*Color****			
Grayscale color	4 (9.5%)	42 (93.3%)	14 (16.9%)
Classic dark	8 (19.0%)	3 (6.7%)	10 (12.0%)
Classic bright	19 (45.2%)	0 (0.0%)	34 (41.0%)
Unique color	11 (26.2%)	0 (0.0%)	25 (30.1%)
*Pattern****			
No pattern	4 (9.5%)	14 (31.1%)	9 (10.8%)
Stripe and check	8 (19.0%)	11 (24.4%)	39 (47.0%)
Obscure pattern	16 (38.1%)	11 (24.4%)	10 (12.0%)
Other patterns w/o Stripe and check	14 (33.3%)	9 (20.0%)	25 (30.1%)
*Functionality****			
Do not care about function	0 (0.0%)	2 (4.4%)	6 (7.2%)
Protection	22 (52.4%)	3 (6.7%)	48 (57.8%)
Comfort	12 (28.6%)	19 (42.2%)	28 (33.7%)
Easy care	8 (19.0%)	21 (46.7%)	1 (1.2%)
*Composition****			
Do not care about composition	1 (2.4%)	0 (0.0%)	0 (0.0%)
Natural fibers	41 (97.6%)	13 (28.9%)	29 (34.9%)
Contain man-made fibers	0 (0.0%)	32 (71.1%)	54 (65.1%)

Note ***$p < 0.001$ for chi-square tests among the three clusters

Distinctions Among the Three Clusters in Terms of Fabric Features

As indicated in Table 4 and Fig. 2, the three clusters differed significantly from each other in terms of consumers' preferred fabric features for high-end shirts.

The distinctions among the three clusters in terms of fabric features are depicted visually in Fig. 3.

In terms of color, the majority of participants in Cluster 1 and Cluster 3 preferred bright color or unique color, while no single participant in Cluster 2 chose any of these two color options. Instead, a dominant number of Cluster 2 members (93.3%) preferred grayscale colors.

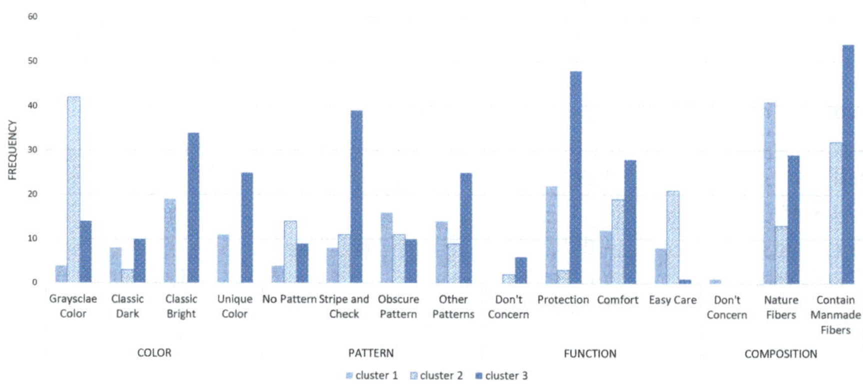

Fig. 2 Fabric feature preference among three clusters displayed in bar chart

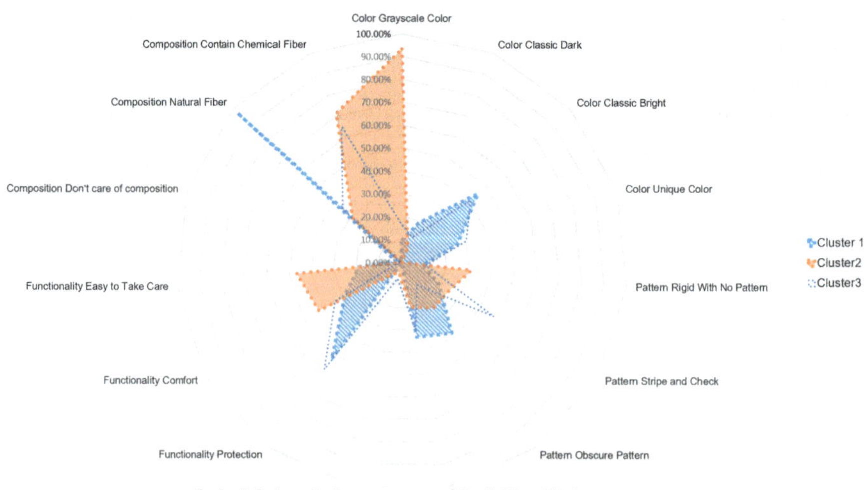

Fig. 3 Distinctions among the three clusters in terms of fabric attributes

In terms of pattern, while the preference was relatively evenly distributed among the four types of pattern options for Cluster 2, there was a clear division among both Cluster 1 and Cluster 3. About 70% of Cluster 1 participants preferred obscure pattern or patterns without check/stripes. For Cluster 3, about half of the participants preferred check/stripe pattern, with about a quarter of the participants preferred patterns without check/stripes.

In terms of functionality, while protection was viewed as important for over half of Cluster 1 and Cluster 3 members, only less than 10% of Cluster 2 members deemed protection important. Instead, majority of Cluster 2 members valued comfort and easy to care functions. Comfort was viewed as important for a modest group in both Cluster 1 and Cluster 3.

In terms of composition/materials, Cluster 1 members dominantly (97.6%) preferred natural fibers. However, Cluster 2 and Cluster 3 each had about 70% members who would accept blends that contain man-made fibers.

High-End Shirt Consumption Behaviors Among the Three Clusters

A cross-tabulation was made among the three clusters in terms of their high-end shirt consumption behaviors (Table 5). The three clusters were very similar to each other in terms of purchase experience of shirts above CNY500 (USD78) and preferred retail channels for high-end shirts. Across the three clusters, over half of the respondents had purchased shirts priced above CNY500 (USD78). About 70% of

Table 5 High-end shirt consumption behaviors among the three clusters

	Cluster 1 (n = 42, 24.7%)	Cluster 2 (n = 45, 26.5%)	Cluster 3 (n = 83, 48.8%)
Purchase experience of shirts priced over CNY500 (USD78)			
No	16 (38.1%)	17 (37.8%)	36 (43.4%)
Yes	26 (61.9%)	28 (62.2%)	47 (56.6%)
Purchase frequency			
Monthly	4 (9.5%)	8 (17.8%)	7 (8.4%)
Quarterly	12 (28.6%)	9 (20.0%)	11 (13.3%)
Less often than quarterly	26 (61.9%)	28 (62.2%)	65 (78.3%)
Distribution channel			
Brand boutique	29 (69.0%)	34 (75.6%)	64 (77.1%)
Bespoke store	8 (19.0%)	4 (8.9%)	9 (10.8%)
Online store	5 (11.9%)	7 (15.6%)	10 (12.0%)
*Purchase intention of shirts priced over CNY1,000 (USD150)**			
No	11 (26.2%)	16 (35.6%)	42 (50.6%)
Yes	31 (73.8%)	29 (64.4%)	41 (49.4%)

Note *$p < 0.01$ for chi-square tests among the three clusters

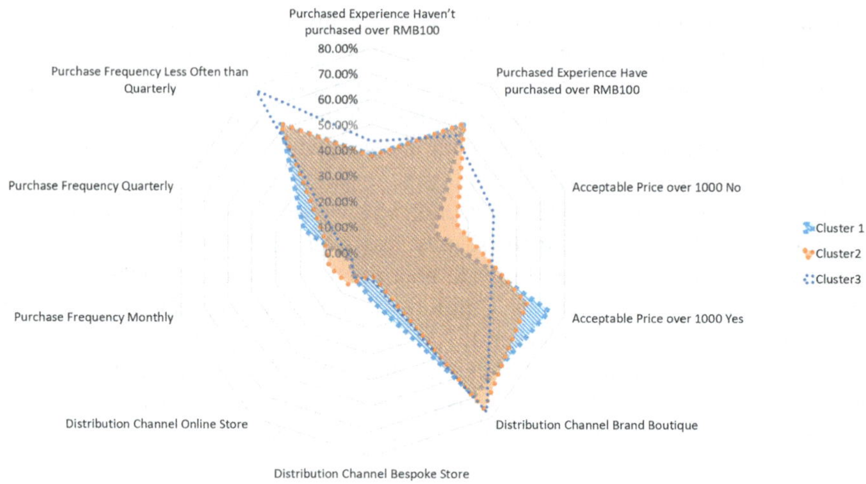

Fig. 4 High-end consumption behaviors of the three clusters

the participants in each of the three clusters indicated their preference of purchasing high-end shirts from specialty retailers/brands over bespoke or online stores. While it was not significant, light difference was observed among the three clusters in terms of their purchase frequency of high-end shirts. In general, participants in all three clusters did not purchase high-end shirts frequently. More than 80% of Cluster 3 participants purchased less often than quarterly, while the percentage was only about 60% for the other two clusters. Clusters 1 and 2 had more people shopped either monthly or quarterly in comparison to Cluster 3. A significant difference was identified among the three clusters in terms of participants' intention to purchase shirts priced above CNY1,000 (USD150). While the majority of Cluster 1 and Cluster 2 consumers would like to buy shirts priced over CNY1,000 (USD150), Cluster 3 was divided equally between the 'yes' and 'no' intention groups. Figure 4 visually depicts the high-end shirt consumption behaviors of consumers among the three clusters.

Demographic Profiles for the Three Clusters

The cross-tabulations (Table 6) on the demographics of the three clusters do not suggest significant differences. The three clusters were quite similar in their demographic profiles. About 60% of the participants for all three clusters were in the age range of 25–45, with about 30% above 45. In terms of monthly income, interestingly, across the three clusters, about 70% of the consumers had less than USD2,300 and 20% had more than USD3,600, with the remaining 10% in between. The education distribution was also very similar among the three clusters, with 85% of the respondents had at least some college or above. About 20–30% of

Table 6 Demographic profiles for the three clusters

Demographics	Cluster 1 (n = 42, 24.7%)	Cluster 2 (n = 45, 26.5%)	Cluster 3 (n = 83, 48.8%)
Age			
18–24	5 (11.9%)	4 (8.9%)	2 (2.4%)
25–34	13 (31.0%)	22 (48.9%)	39 (47.0%)
35–44	10 (23.8%)	6 (13.3%)	12 (14.5%)
45–54	6 (14.3%)	10 (22.2%)	18 (21.7%
over 55	8 (19.0%)	3 (6.7%)	12 (14.5%)
Monthly income (USD)			
Less than 1,300	16 (38.1%)	22 (48.9%)	34 (41.0%)
1,300–2,300	15 (35.7%)	10 (22.2%)	24 (28.9%)
2,300–2,700	3 (7.1%)	3 (6.7%)	3 (3.6%)
2,700–3,600	0 (0.0%)	1 (2.2%)	6 (7.2%)
Over 3,600	8 (19.0%)	9 (20.0%)	16 (19.3%)
Education background			
High school diploma or less	6 (14.3%)	7 (15.6%)	10 (12.0%)
College	23 (54.8%)	23 (51.1%)	54 (65.1%)
Advanced degree	13 (31.0%)	15 (33.3%)	19 (22.9%)
Oversea background			
No	32 (76.2%)	32 (71.1%)	66 (79.5%)
Yes	10 (23.8%)	13 (28.9%)	17 (20.5%)

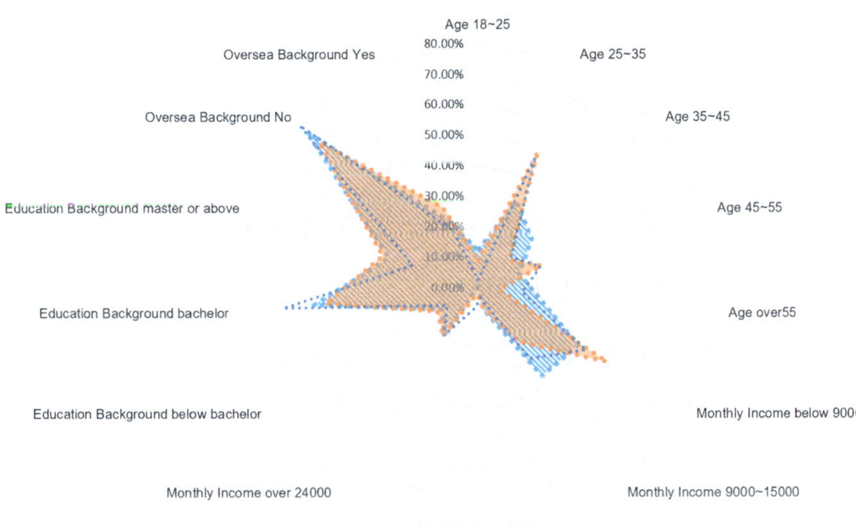

Fig. 5 Demographic profiles of the three clusters

respondents in each cluster had overseas experiences. Figure 5 depicts the demographic profiles of the three clusters.

Conclusions

With rising disposable incomes and male consumers' increasing interest in personal grooming products, the Chinese menswear market is growing tremendously. Shirt is one of the most popular menswear items chosen by Chinese male consumers for both casual and formal occasions. While the shirt market is growing, there are few shirt fabric brands in the Chinese market. At the same time, while many Chinese fabric manufacturers possess great capacity in producing high-end fabrics and supply to globally known upscale shirt brands, no fabric brand is in existence in China for high-end shirts. Realizing that ingredient (fabric) branding can bring additional value to the final product (shirts), especially for upscale products, it might be to the great benefits of the current fabric enterprises to brand their fabric products. Consumer insights of the high-end men's shirt market in China are very critical, but lacking. The current study was designed to fulfill this gap.

Based on consumers' preference of fabric features for high-end shirts, three distinctive clusters were identified in this study. Consumers in Cluster 1 had a high intention to purchase high-end shirts and developed a taste and an appreciation of fashion in their high-end shirt consumption. Consumers in Cluster 2 exhibited a pragmatic orientation by focusing on the easy-care function of the shirts and choosing online shopping. Fashion seemed not so important to these pragmatic consumers as they favored grayscale colors and plain no-pattern fabric. Man-made fibers were most accepted by these consumers. Consumers in Cluster 3, like those premium consumers in Cluster 1, displayed a strong taste and appreciation of fashion through their preference in fabric features. However, these consumers did not have the same purchase frequency and purchase intention for high-end shirts compared to the premium consumers. The following is a detailed characterization for each cluster identified in the study.

Cluster 1: Premium Consumers

About a quarter of the respondents belonged to this cluster of premium consumers. The members of this cluster shared a similar preference of high-end shirts in terms of color, pattern, materials, and functions. Fashion seemed to be valued by these premium consumers. They preferred bright and unique colors and more sophisticated patterns such as obscure patterns and patterns without traditional stripe and check components. Natural fibers were extremely important to these consumers. In terms of function, they preferred protection and comfort. But, in contrast, the attribute of easy care seemed not important for these consumers. These consumers also had a significantly much higher intention in purchasing high-end shirts priced above CNY1,000 (USD150). Similar to the other two clusters, consumers in this cluster mainly shopped at special stores for high-end shirts. While the difference was not significant, a relatively higher percentage of consumers in this cluster used bespoke to fulfill their needs for high-end shirts than consumers in the other two clusters. In terms of demographics, this cluster had a high percentage of members holding a more advanced degree. Also, this cluster seemed having a relatively heavier concentration on the age groups of 35–45 and over 55. While the income distributions seemed not so different among the three clusters, this cluster relatively had a heavier concentration of high-income consumers.

Cluster 2: Pragmatic Consumers

This cluster took up another quarter of the sample. The consumers in this cluster manifested a high pragmatic orientation. Fashion was not viewed as important to these consumers as to the premium consumers. A dominant percentage of these consumers preferred grayscale colors with no one showing interest in bright or unique colors. Worth noting, a high percentage of consumers in this cluster preferred no-pattern plain shirts. The preference for other patterns was equally distributed among the members. Another manifestation of these consumers' pragmatic orientation was their strong attention to comfort and the easy-care function of the shirts. And relevantly, these consumers were agreeable to have their shirts made of man-made fibers. While not as many as in the first cluster, over 60% of consumers in this cluster indicated interests in purchasing high-end shirts priced above CNY1,000 (USD150). The retail channels used to high-end shirts were pretty similar to the other two clusters, but with a relatively heavier usage of online stores. Comparing to the other two clusters, consumers in this cluster purchased high-end shirts more frequently. In terms of demographic, the education level was pretty similar to the first cluster with a high percentage of consumers holding an advanced degree.

Cluster 3: Economic Consumers

Around half of the respondents fell into Cluster 3. These consumers indicated the weakest intention to purchase high-end shirts priced above CNY1,000 (USD150) and purchased high-end shirts least frequently. A high percentage of consumers in this cluster could accept man-made fiber materials for high-end shirts. While being economic, these consumers paid similar attention to fashion as the premium consumers did, in that they favored bright or unique colors and functions of protection and comfort. However, these consumers either loved or hated the traditional stripe and/or check patterns. They had similar preference to consumers in the other two clusters in terms of shopping channels for high-end shirts, but with relatively more patronage at special stores. In terms of demographics, again no significant difference was identified among the groups. However, this cluster had a relatively heavier concentration on college education, while the other two clusters had more members holding a more advanced degree.

Among the three clusters, the economic consumers cluster represented the largest group including about half of the respondents. These consumers had a keen interest in and taste for high-end shirts, but their purchase intention was the lowest for shirts priced above CNY1,000 (USD150) and their high-end shirt purchase was the least frequent. In comparison, the premium consumers and pragmatic consumers were more likely to be the target market for high-end shirts in terms of their purchase experience and intention. However, much difference was found between these two groups in their shirt consumptions with one fashion oriented and the other utilitarian oriented reflected by their fabric attribute preference. These findings can provide great consumer insights for the high-end shirt brands as well as fabric brands in their product development and marketing endeavors serving the Chinese market.

Limitations and Future Research

While very insightful, the study bears several limitations. First limitation comes from the small convenience sample. Therefore, the results need to be generalized with caution. Secondly, the three clusters did not distinguish from each other significantly in terms of demographics. The limitation from the convenience sample might have partially contributed to the insignificance. But, future research could also explore the possibility of difference among the segments in terms of other variables, such as lifestyle, personality, social class. Also, future studies could investigate Chinese male consumers' high-end shirt consumption behaviors from other perspectives, such as brand loyalty, country of origin influence, fit/style preference, and consumption motivations.

References

Albini. (2017). *Albini official website*. Accessed October, 2017 online at: https://www.albinigroup.com/.

Alumo. (2017). *Alumo official website*. Accessed October, 2017 online at: https://www.alumo.ch/en.

Bakewell, C., Mitchell, V. W., & Rothwell, M. (2006). UK generation Y male fashion consciousness. *Journal of Fashion Marketing and Management: An International Journal, 10* (2), 169–180.

Banerjee, S., & Agarwal, N. (2013). Cluster and conjoint analysis of the entry level brands in casual wear market in India w.r.t. product attributes. *Journal of Marketing & Communication, 8*(4).

Bizvibe. (2015). Global menswear market growth shows no signs of slowing. Accessed October, 2017 online at: http://www.bizvibe.com/blog/global-menswear-market-growth-shows-no-signs-slowing/.

Business of Fashion. (2014). *No signs of slowing in the global menswear market*. Accessed October, 2017 online at: http://www.businessoffashion.com/articles/fashion-show-review/signs-slowing-global-menswear-market.

Canclini. (2017). *Canclini official website*. Accessed October, 2017 online at: https://www.canclini.it/.

China Business Industry Research. (2016). *Prospect report of 2017 China formal menswear market size*. Accessed October, 2017 online at: http://www.askci.com/news/chanye/20161216/10335783176.shtml (in Chinese).

China Business Industry Research. (2017a). *Analysis report of 2017 China menswear industry*. Accessed October, 2017 online at: www.askci.com/news/chanye/20170802/103450104420.shtml (in Chinese).

China Business Industry Research. (2017b) *The size and development prospect of China's menswear market in 2017*. Accessed October, 2017 online at: www.askci.com/news/life/20161216/10225283169.shtml (in Chinese).

China Luxury Advisors. (2015). *China luxury advisor's thought leadership*. Accessed October, 2017 online at: http://chinaluxuryadvisors.com/chinas-menswear-market-shifts-into-high-gear/.

China Textile Planning Research Association. (2014). *Analysis on the overall characteristics of China's textile and apparel listed companies and their investment value*. Accessed November, 2017 online at: http://www.ctpra.com/hyzx/hyxw/201409/t20140926_1844699.html (in Chinese).

Edited. (2016). *Spring 2017 menswear: Market growth & trends*. Accessed October, 2017 online at: http://edited.com/blog/2016/06/menswears-magic-moment/.

Elliot, A. J., & Niesta, D. (2008). Romantic red: Red enhances men's attraction to women. *Journal of Personality and Social Psychology, 95*(5), 1150.

EFU. (2017). *"2017 China men's wear industry development report" released, which showed greater competition*. Accessed October, 2017 online at: http://news.efu.com.cn/newsview-1234248-1.html.

Euro Monitor. (2017a). *Apparel and footwear global industry overview*. Accessed October, 2017 online at: http://www.portal.euromonitor.com.prox.lib.ncsu.edu/portal/analysis/tab.

Euro Monitor. (2017b). *Menswear in Western Europe*. Accessed October, 2017 online at: http://www.portal.euromonitor.com.prox.lib.ncsu.edu/portal/analysis/tab.

Euro Monitor. (2017c). *Menswear in the US*. Accessed October, 2017 online at: http://www.portal.euromonitor.com.prox.lib.ncsu.edu/portal/analysis/tab.

Euro Monitor. (2017d). *Menswear in China*. Accessed October, 2017 online at: http://www.portal.euromonitor.com.prox.lib.ncsu.edu/portal/analysis/tab.

Fashion United. (2016). *Menswear will "spearhead" growth in clothing market*. Accessed October, 2017 online at: http://fashionunited.uk/news/retail/menswear-will-spearhead-growth-in-clothing-market/2016072721202.

Frith, H., & Gleeson, K. (2004). Clothing and embodiment: Men managing body image and appearance. *Psychology of men and masculinity, 5*(1), 40–48.

Gentleman's Gazette. (2010). *Shirt basics guide: Material & fabric*. Accessed October, 2017 online at: https://www.gentlemansgazette.com/shirt-guide-basics-weaves.

Gentleman's Gazette. (2013). *The dress shirt guide-making & hallmarks of a quality shirt*. Accessed October, 2017 online at: https://www.gentlemansgazette.com/dress-shirt-guide-making-hallmarks-quality.

Getzner. (2017). *Getzner official website*. Accessed October, 2017 online at: https://www.getzner.at/en/.

Giuseppe. (2014). *Annual financial report*. Accessed October, 2017 online at: http://money.finance.sina.com.cn/corp/view/vCB_AllBulletinDetail.php?stockid=002687&id=1691805 (in Chinese).

Giuseppe. (2015). *Annual financial report*. Accessed October, 2017 online at: http://money.finance.sina.com.cn/corp/view/vCB_AllBulletinDetail.php?stockid=002687&id=2284193 (in Chinese).

Giuseppe. (2016). *Annual financial report*. Accessed October, 2017 online at: http://money.finance.sina.com.cn/corp/view/vCB_AllBulletinDetail.php?stockid=002687&id=3246362 (in Chinese).

Guokr. (2014). *Standard of high quality shirts*. Accessed October, 2017 online at: https://www.guokr.com/post/562905 (in Chinese).

Hart, C., Farrell, A. M., Stachow, G., Reed, G., & Cadogan, J. W. (2007). Enjoyment of the shopping experience: Impact on customers' repatronage intentions and gender influence. *The Service Industries Journal, 27*(5), 583–604.

Heilan Home. (2014). *Annual financial report*. Accessed October, 2017 online at: http://money.finance.sina.com.cn/corp/view/vCB_AllBulletinDetail.php?stockid=600398&id=1639492 (in Chinese).

Heilan Home. (2015). *Annual financial report*. Accessed October, 2017 online at: http://money.finance.sina.com.cn/corp/view/vCB_AllBulletinDetail.php?stockid=600398&id=2406054 (in Chinese).

Heilan Home. (2016). *Annual financial report*. Accessed October, 2017 online at: http://money.finance.sina.com.cn/corp/view/vCB_AllBulletinDetail.php?stockid=600398&id=3102059 (in Chinese).

Gaskin, J. (2012). *Two step cluster analysis in SPSS*. Accessed October, 2017 online at: https://www.youtube.com/watch?v=DpucueFsigA.

Mintel. (2016). *Men's clothing*. Accessed October, 2017 online at: http://reports.mintel.com/display/747672/.

Mintel. (2017). *Menswear market becomes more competitive*. Accessed October, 2017 online at: http://academic.mintel.com/Sinatra/oxygen/print/id=840853.

Monti. (2017). *Monti official website*. Accessed October, 2017 online at: https://www.monti.it/.

Niesta Kayser, D., Elliot, A. J., & Feltman, R. (2010). Red and romantic behavior in men viewing women. *European Journal of Social Psychology, 40*(6), 901–908.

Nippon Color & Design Research Institute Inc. (2017). *Image scale system*. Accessed October, 2017 online at: http://www.ncd-ri.co.jp/english/main_0104.html.

Osgerby, B. (2003). A pedigree of the consuming male: masculinity, consumption and the American 'leisure class'. *The Sociological Review, 51*(S1), 57–85.

Pinpai China. (2017). *R&D of private customized system to promote the transformation of enterprise*. Accessed October, 2017 online at: http://pinpai.china.com.cn/yxl/2017-04/18/content_9440894.htm (in Chinese).

Prnewswire. (2016). *Global menswear market to grow at a CAGR of 4.51% till 2020*. Accessed October, 2017 online at: http://www.prnewswire.com/news-releases/global-menswear-market-to-grow-at-a-cagr-of-451-till-2020-300329086.html.

Rinallo, D. (2007). Metro/fashion/tribes of men: Negotiating the boundaries of men's legitimate consumption. *Consumer tribes*, 76–92.

Ross, F. (2007). Refashioning London's bespoke and demi-bespoke tailors: New textiles, technology and design in contemporary menswear. *Journal of the Textile Institute, 98*(3), 281–288.

Rundle-Thiele, S., Kubacki, K., Tkaczynski, A., & Parkinson, J. (2015). Using two-step cluster analysis to identify homogeneous physical activity groups. *Marketing Intelligence & Planning, 33*(4), 522–537.

Saint Angelo. (2014). *Annual financial report*. Accessed October, 2017 online at: http://finance.sina.com.cn/corp/view/vCB_AllBulletinDetail.php?stockid=002154&id=1680304 (in Chinese).

Saint Angelo. (2015). *Annual financial report*. Accessed October, 2017 online at: http://money.finance.sina.com.cn/corp/view/vCB_AllBulletinDetail.php?stockid=002154&id=2389108 (in Chinese).

Saint Angelo. (2016). *Annual financial report*. Accessed October, 2017 online at: http://money.finance.sina.com.cn/corp/view/vCB_AllBulletinDetail.php?stockid=002154&id=3335442 (in Chinese).

Sarstedt, M., & Mooi, E. (2014). Cluster analysis. In *A concise guide to market research* (pp. 273–324). Berlin, Heidelberg: Springer. Retrieved from: http://www.bookmetrix.com/detail/chapter/0a55d9dc-a197-4ffa-ba79-362a8db95023#downloads.

Septwolves. (2014). *Annual financial report*. Accessed October, 2017 online at: http://money.finance.sina.com.cn/corp/view/vCB_AllBulletinDetail.php?stockid=002029&id=1692878 (in Chinese).

Septwolves. (2015). *Annual financial report*. Accessed October, 2017 online at: http://money.finance.sina.com.cn/corp/view/vCB_AllBulletinDetail.php?stockid=002029&id=2312955 (in Chinese).

Septwolves. (2016). *Annual financial report*. Accessed October, 2017 online at: http://money.finance.sina.com.cn/corp/view/vCB_AllBulletinDetail.php?stockid=002029&id=3177240 (in Chinese).

Shaw, P., & Tan, Y. (2014). Race and masculinity: A comparison of Asian and Western models in men's lifestyle magazine advertisements. *Journalism & Mass Communication Quarterly, 91*(1), 118–138.

Soktas. (2017). *Soktas official website*. Accessed October, 2017 online at: https://www.soktas.com.tr/.

Storey, N. (2008). *History of men's fashion* (pp. 99–104). South Yorkshire, Great Britain: Remember When Publishing.

Testa. (2017). *Testa official website*. Accessed October, 2017 online at: https://www.testaspa.com/site/.

Wrenn, B., Steven, R. E., & Loudon, D. L. (2002). *Text and cases* (2nd ed.). New York: The Haworth Press. Retrieved from: https://zh.scribd.com/doc/102616412/Marketing-Research-Text-and-Cases.

Youngor. (2014). *Annual financial report*. Accessed October, 2017 online at: http://money.finance.sina.com.cn/corp/view/vCB_AllBulletinDetail.php?stockid=600177&id=1769289 (in Chinese).

Youngor. (2015). *Annual financial report*. Accessed October, 2017 online at: http://money.finance.sina.com.cn/corp/view/vCB_AllBulletinDetail.php?stockid=600177&id=2439244 (in Chinese).

Youngor. (2016). *Annual financial report*. Accessed October, 2017 online at: http://money.finance.sina.com.cn/corp/view/vCB_AllBulletinDetail.php?stockid=600177&id=3367452 (in Chinese).

Chinese Consumers and the Knitwear Fashion Market

Xin Zhang, Colin Gale and Claudia Eckert

Abstract The Chinese knitwear industry is in crisis and needs to adapt to survive. Chinese knitwear brands face the problems of a shrinking domestic market, an aging consumer base, and lack of appeal to new generations. Meanwhile, products created by the industry are becoming more homogeneous. Influenced by globalization, Chinese consumers have a wide choice of products with different styles and brands in the fashion market. As more and more international brands enter into the Chinese market, offering clothes that appeal to younger consumers, the challenge for many established Chinese fashion brands, especially in knitwear, is increasing. This paper reports on research into creative design in Chinese knitwear brands, which revealed that the Chinese knitwear industry is lacking a creative culture. In many knitwear companies in China, the lack of understanding of the value of design stems from a number of factors: restricted creative freedom for designers in companies and an absence of practice in the design education system for creative processes to develop. This results in companies lacking confidence in their designers' creative abilities. This paper introduces the Chinese knitwear market and fashion consumers in China. It discusses why many Chinese consumers prefer luxury or international brands rather than domestic brands and analyses the gaps in creative design between Chinese and successful international brands. This is based on research into consumer behavior for knitwear. It concludes by making recommendations on how the industry can turn things around.

Keywords Chinese fashion market · Knitwear design · Consumer behaviors · Design culture · Consumption trend

---

X. Zhang (✉) · C. Gale
Birmingham City University, Birmingham, UK
e-mail: Xin.Zhang8@mail.bcu.ac.uk

C. Gale
e-mail: Colin.Gale@bcu.ac.uk

C. Eckert
Open University, Milton Keynes, UK
e-mail: claudia.eckert@open.ac.uk

Y. Xu et al. (eds.), *Chinese Consumers and the Fashion Market*, Springer Series in Fashion Business, https://doi.org/10.1007/978-981-10-8429-4_9

Introduction

China has grown into an important market after the reforms and opening policies, which began in the 1980s. Income, quality of life, and consumption power of Chinese people have rapidly improved with the growing Chinese economy. Many international fashion brands, including numerous luxury and fast fashion brands such as Gucci, Louis Vuitton, and Zara, have opened stores in China with local Chinese brands also becoming established. In the knitwear industry, two types of knitwear brands play an important role in the Chinese retail market: Merino wool garments, mostly located in Zhejiang province, and cashmere knitwear and related products from Inner Mongolia, Zhejiang Province and Shenzhen city. Chinese knitwear has traditionally occupied large-sale areas in department stores in China with strong sales performances. Since their peak around the year 2000, many Chinese knitwear brands faced a shrinking market and aging consumers (Anon, 2008). While cashmere knitwear brands are still represented in many department stores, it has become almost impossible to find merino knitwear brands in department stores in China. But, where there are knitwear products, they have traditional and conservative styles and the designs are very similar to each other. This undermines competitiveness and does not attract a new generation of customers. This research focuses on the problems in Chinese knitwear design culture and brand strategies and aims to identify ways to support the knitwear companies to improve their brands and bring back their market share. The findings are based the first author's experience as an industrial knitwear designer in China from the late 1990s to the present, as well as interviews with other designers and managers in knitwear companies.

In the first part of Section "Background", the problems of the Chinese knitwear sector are examined by reviewing the past and current state of the industry and the changes that have occurred. This then explains how consumers attitudes have changed and the rise in lifestyle and consumption has affected traditional retail systems, including counterfeiting. Different types of company behaviors are introduced in the final part of this section. Section "Methodology" introduces the research methodology. Due to the issues, the knitwear industry needs to respond, but faces its own challenges, particularly in brand identity, poor market structure and weak design culture which is explained in Section "The Challenges of the Chinese Knitwear Industry". This is in part caused by limitations in design education in China. The recommendations and suggestions will be discussed in Section "Recommendation for Changing the Situation". Finally, it is concluded that the Chinese knitwear industry is in crisis and needs to adapt to survive.

Background

Knitwear must be seen in the wider context of fashion in China. This requires understanding both the brands and the Chinese customers.

Knitwear Industry in China

China has become a more dynamic and significant market in the last few decades. The continually growing economy has increased income for a significant number of people. In particular, middle class spending on luxury goods has been expanding. For example, Chinese consumers spent around $116.8 billion on luxury goods in 2015, which represents 46% of luxury consumption worldwide (Anon, 2016). On average, this market increased by around $6 billion per year between 2009 and 2015 (Li, Li, & Kambele, 2012).

Global Versus Domestic Brands

Many famous global brands have their own flagship stores in China, occupying the prime locations in commercial centers in the more economically developed cities. For instance, in Guangzhou, the third largest Chinese city, more than ten shopping centers with the best positions given to famous high-class international brands, to attract customers into shopping areas or shopping malls. Although the first Chinese fashion company was founded in 1917 in Shanghai (Tsui, 2010), it was not until the reform in the 1980s that domestic brands grew and expanded. After this opening up of China, national brands gradually developed through the 1980s and 1990s (Li, 2004). Many brands developed quality products leading to success in the Chinese market back then. Interestingly, there were over 5000 knitwear companies and around 180 brands in China in the 1990s (Yang, 2009). Most knitwear companies were running their businesses as knitwear suppliers (5000) rather than brands (180). After that, more and more knitwear companies began to run their own labels. Some of them developed and became large businesses by the early 2000s, covering most Chinese provinces. Some brands such as the cashmere brand Erdos had over 1500 stores at their peak time. However, there is now a different trend in the Chinese fashion market.

Brand Proliferation in the Market

There are thousands of clothing brands in the Chinese retail system, and new brands appear in the market every year, while others disappear. For example, the authors found that some knitwear brands had hundreds of millions US$ annual volume sales, but profits were low, just covering their costs. This was due to a high percentage of discounted products. The situation even resulted in setting up a new business for selling inventory goods. In 2012, the annual business sales in the Chinese biggest inventory market from Yiwu city Zhejiang province were over 5 billion US$ (Chen, 2013).

Some famous Chinese brands have recently closed most of their retail stores. These companies used to be developing quickly and expanded in a short period without analyzing their market demand. They did not have enough experience to run a successful brand, which resulted in too many stores with high expenditure and many unpopular clothes to sell. The companies also compete with other brands with very similar product offerings. For example, to occupy a greater market share, some cashmere companies were running three or four brands with slightly different yarns, but similar designs. However, most of these brands did not further develop as they expected. In fact, too many indistinctive brands proved unattractive as consumers become exposed to global brands. Unsold stock is kept in the stores at the end of the season and discounted, leaving many Chinese brands with surprisingly large stocks (Chen, 2013). Most knitwear brands try to relieve their inventory pressure by keeping discounted products on sale throughout the year rather than discounting at certain time of the year. This points to a lack of good system for both market forecast and management of their products.

Decline of Chinese Brands

Chinese fashion brands have developed quickly over the past four decades. Many of them have developed from small factories with a few employees to big companies with hundreds of chain stores during their peak time. However, there has been a decline in sale performance facing the domestic brands in recent years in the Chinese fashion market. Since 2012, many large Chinese brands have had to close large numbers of their shops (Zhang, 2016). For example, Chinese sports brands closed over 3000 stores between 2012 and 2014, and Bosideng closed around 5000 stores up to 2015, which all reflect the problems arising both from economic slowdown and an unsustainable rapid expansion. Chinese knitwear brands that are specialized in a single type of product suffered more than diversified fashion brands. Chinese wool knitwear brands almost disappeared in most big cities, and many small- and middle-scale cashmere brands went out of business between 2000 and 2010. For instance, in 2008, there were 718 Merino woolen knitwear brands with 2.6% of salerooms and 295 cashmere knitwear brands with 8% of salerooms in Beijing's fashion market (Anon, 2009). Now, there are only around 10 big cashmere knitwear brands left in Beijing. Most of the declining knitwear brands do not have clear and strong brand personalities; their products are easily replaced by other similar products with cheaper prices. The situation reflects a trend for oversupply in the Chinese fashion market leading to increased competition and brand failure (Liu, 2007).

Fashion Consumers in China

Lifestyle and Value Changes

China is attracting the world's attention with its rapid economic growth. The quality of life is rising universally across the country with improvements in the standard of living and lifestyle changes (Xu, 2013). Chinese people were used to buying clothes just to satisfy their basic needs, such as keeping warm. After consumers' income increased in the last few decades their consumption style has changed. People use clothes to express their style and individuality as well as their status and positions in their social life (Xu, 2013). According to statistics, by the end of 2013, China has become the world's largest consumer market for luxury goods (Sun, 2014). "Face consciousness" is an important cultural aspiration for the Chinese society. The "face" means personal need of sense of social values reflecting from other members in society, which is also reflected by people's fashion consumptions and shopping behaviors nowadays (Chen, 2011). Face culture is part of traditional Chinese shame culture, which is influenced by Confucianism; people are keen on "seeking self-perfection" and "fear of loosing face," and "the feeling state after one's face has been lost" (Zhang, 2015). Therefore, many people care about the feelings others have toward them and the feedback they get on their appearance including their clothes, because this is related to their "face." This influences their personal image, social status, wealth, and social networks (Chen, 2011). Thus, most Chinese consumers look for social identities through their clothes and quality of life, which led to many pursuing famous brands and comparing themselves to others in the Chinese society (Chen, 2011). Chinese people, especially those living in large cities, adore luxury lifestyles and prefer showing their success to others by wearing or using high-price products. Influenced by this "face" culture, many people use expensive products not because of the products' function but to meet a psychological need, conform to peer pressure, or sometimes just to draw other people's attention. This is particularly true for young people (Zhao, 2009). For example, in recent years, many college students are using parents' money to buy luxury goods. Some of them are not from rich families but still ask money from their parents, and some students even borrow money from others just to satisfy their consumption desires (Anon, 2017).

Consumer Groups and Consumption Preferences

Many Chinese fashion brands are keen to segment target consumers by age, as each generation has grown up in a totally different situation and has different educational backgrounds. According to Euromonitor International (2017) "young adults" (people born from 1980s to 2000s) care about individual identity; the "middle youth" (1960s–1990s) care more about balancing their lives and jobs; the "mid lifers" (1940s 1970s) have strong purchasing power and like luxury brands; and the "later-lifers" (1920s–1960s) make health and healthcare a big priority for them. Wang (2014) divided the main Chinese fashion consumers into three age groups, which includes people born during 1970s or 1980s or 1990s."Most of 70s consumers have temperance and cherish what comes to them"; Chinese of the 80s received better education and high-quality culture, along with globalization and development of the Internet, and they have formed their own characteristic of pursuing personality and innovation without being restrained by traditions; "For the 90s, they are a fast-growing consuming group; they are not just individuals consuming, but also have a great influence on family purchasing" (Wang, 2014). Chinese born in the "80s and 90s are becoming the core consumer group, while most of 70s have been the backbone force of the society" (Wang, 2014). Compared with young consumer fashion brands, there are few brands designed for middle-aged consumers. Normally, there are two opposite types of brands for middle age groups; one offering old-fashioned clothes with low qualities and prices, and another offering higher quality with expensive prices. This does not leave enough choices for middle-aged consumers for fashionable clothes with reasonable prices in Chinese fashion markets (Yu, 2015). Based on different income and personal background Yu (2015) identified the following five types of consumption types: free and simple; indifferent; following fashion; style conscious; and conservative. The elderly are keen to buy clothes for their basic physiological needs, but care about the price of the clothes (Lao, 2013). Consumers can also be divided by cities and areas. There are four tiers of cities with different development levels in China. Generally, people living in the first-tier cities are more influenced by cultural globalization and are quicker to follow international trends, while people in third-tier cities or remote areas are more conservative. For example, many Shanghai people prefer to buy fashion products when they travel to other countries (Ye, 2013), which mean their fashion tastes are closer to international brands.

China is a large country with 56 different ethnic groups in different regions. The different ethnic groups or areas have their own fashion cultures and aesthetical standards. This reflects their lifestyles and different preferences in clothes styles and colors. For example, one knitwear brand has always sold clothes with different colors to different provinces in China. Thus, if knitwear brands want to meet consumers' expectations, they have to build clear brand and market positioning.

Traditional In-store Retailers

The traditional in-store retailers face many challenges from changes in the Chinese market. Many Chinese city people, with busy jobs or care duties, do not have enough time to go to shops after work or on weekends. Many Chinese people are now buying goods in online stores, because they are offered at cheaper prices with fast delivery and convenience. Thus, over 97% of city consumers think about online shopping first when they need to buy something (Anon, 2013). This puts traditional retailers under pressure, especially the traditional department stores. During 2012 to 2014, over 200 department stores were closed in China (Situ, 2015). Firstly, fewer people now visit traditional department stores, because customers prefer shopping malls with entertainment and leisure facilities as it has become popular to merge shopping and leisure activities. Secondly, traditional department stores offer many brands with similar products without distinguishing characteristics. For example, many department stores in China used to devote an entire floor to knitwear brands, but have now changed to shopping mall style arrangements where all the items of one brand are kept together. These changes have caused many Chinese knitwear brands to disappear from many cities, because of their single product structure and old-fashion styles. At the same time, the increase in rental and staff costs has also caused the profits of department stores to decline from around 7–8% per year to 2 and 3% (Lin, 2012).

Over Consumption

Over consumption refers to people buying (much) more than they need. Large numbers of Chinese consumers over consume fashion products (Xu, 2015). This is influenced by Chinese shame culture, "shame is a group rather than an individual concern in China: people almost always belong to a closely integrated group on which their honor or shame is reflected" (Zhang, 2015). Thus, many people subconsciously compare themselves to others clothes and lifestyles within social or friendship groups. In a social group someone might feel shame, if their clothes or appearance are not as good as that of their social peers. Therefore, many Chinese people buy luxury goods although their incomes are not high enough. For example, it is common for a young woman to spend a few months' salary on a luxury bag. In Chinese culture, many people believe that paying attention to pomp and luxury is seen as a symbol of success and status (Liang & Wang, 2007). There is a clear trend toward luxury consumption among younger people, although they do not have enough income to afford the luxury products in China. The average monthly income of a Chinese luxury consumer was 5288 Chinese Yuan (around $768) in 2011. However, the average age of Chinese consumers for the global luxury brands is 15 years younger than in Europe and 25 years less than the US (Qin, 2012). On the other hand, after the fast fashion brands have become popular in China, many people's consumption patterns have changed, because fast fashion brands encourage people to keep following fashion trends by buying fast-changing consumer goods. Consumers buy many clothes to ensure they appear fashionable; however, many garments can just be used for one season because of bad quality and easily outdated designs. Considering these consumption trends, traditional knitwear brands may meet big challenge in the markets, when consumers have many clothes at home for their basic needs. Consumers have high aesthetic requirements when they decide to buy a new item. Therefore, it is opinion of the author that for most of the Chinese knitwear brands, valuable raw materials and functionality are still the selling points rather than design and brand.

Counterfeits

Counterfeit goods refer to a class of illegal goods, which are a complete imitation of brand products in terms of brand name, trademark, function, appearance, and other aspects (Lai & Zaichkowsky, 1999). Compared with the brand products, counterfeit goods can be of inferior quality, but are much cheaper (Lai & Zaichkowsky, 1999). There is a large group of people in China, who aspire to luxury goods, but cannot afford them, which leaves a gap in the market for cheap imitation-luxury goods.

While being illegal in China, some people take the risk and sell counterfeit goods in covert ways because it is a quick way to make money. On the other hand, many consumers do not have the ability to distinguish which products are original or fake (Lai & Zaichkowsky, 1999). In addition, many people are not aware of copyright and buy counterfeits products without being aware that they use an illegal product.

There is little cultural awareness in general of intellectual property rights. In fashion, even if designs are not direct counterfeit they can be very similar in terms of colors, patterns, and details to the designs of other brands. Most Chinese fashion brands prefer to copy parts of designs from famous brands rather than use new designs created by their own design teams. In the knitwear industry, the situation is even more extreme, because a limited number of yarns are used by most knitwear brands leading to greater homogenizations than with fashion products with many kinds of fabric. Knitwear brands also prefer to copy each other to reduce the market risks, despite furious competition and price wars in knitwear markets.

The Behavior of Different Types of Companies

Tradition Knitwear Brands

Traditional Chinese knitwear brands must look for new ways to keep or increase their market shares. For example, one medium-scale cashmere brand that was interviewed in Inner Mongolia in 2015, had over 100 retail stores in China during its peak time, but was now down to just a few. The boss indicated that her company intended to start a new cashmere brand at the end of 2015. Subsequently, the company turned its main business to cashmere goat wool farming. It gave up on cashmere knitwear production, because the brand had lost its market share. Their problems were typical for the Chinese cashmere industry. Despite the quality, knitwear is not designed in collections with a consistent style. Their designs typically fall into two problematic categories:

- Designs with over complicated decorations, that do not match other clothes;
- Designs with too basic old-fashion styles that do not attract buyers.

The Chinese designs stagnate and consumers' aesthetics have developed, influenced by international fashion trends and brands.

Knitwear in Luxury Brands

Chinese consumers prefer luxury clothing and the enjoyment they get out of owning products from international brands with high visibility and reputation (Wang, 2010). Many people believe that it is an outstanding outcome from career success if people can afford luxury products (Wang, 2010). In addition, to attract

Chinese consumers, many luxury brands created new products with specific Chinese cultural elements.

Luxury knitwear can mean knitwear from luxury brands or knitwear made of luxury materials, such as cashmere. Over 75% of the world's cashmere products are manufactured in China. Most luxury knitwear made in China is made by the same companies, who are unsuccessful in maintaining their own brand. However, even with similar quality knitwear made with the same material, there is a total disconnection between luxury international brands and Chinese brands both in terms of price and consumer acceptation. For example, the price range of 100% cashmere knitwear is from $100 to $300 for many Chinese well-known knitwear brands. However, for international luxury brands, similar products double in price. Consumers choose these higher-priced international luxury brands as they show a higher social status. Thus, compared to international luxury brands, Chinese knitwear brands have a long way to go for their brand equity enhancement due to the lack of understanding and strategies for branding and designing.

Methodology

This section reports on a practice-based research study that focused on the problems of design creativity in the Chinese knitwear market. The main observation is that there is a lack of creative culture in the Chinese knitwear industry, including design culture and understanding, branding strategy, and a lack of design education in China. To gain a deep understanding of the situations of the Chinese knitwear brands and find ways for dealing with the problems in knitwear design for these brands, the research methodology was based on a combination of qualitative and quantitative research (see Table 1). This included face-to-face interviews, a case study, design practice research, and online surveys. It also included a comparison between advanced western countries and China in terms of the fashion industries and design education systems to analyze the gap in design education, designers' abilities, and industry environments.

Between 2014 and 2017, face-to-face interviews were conducted with 46 people. The first group of 22 interviewees was from the Chinese and British knitwear industry, and the second group of 24 interviewees was fashion and textile design students and faculty in China, Italy, and the UK. The first author visited five Chinese knitwear companies, comprising two large, two medium-scale, and one small-scale knitwear brands, to talk with their owners and staff during 2015 and 2016. The first author also contacted another big Chinese knitwear brand and talked with the company owner and interviewed a designer from the brand by email in 2016. One British knitwear designer was interviewed in 2016, who had been working for over ten years both in UK and Hong Kong for a British designer studio. A case study, with practice research, was conducted with a Chinese knitwear brand in 2016. The practice research involved designing a knitwear collection for the young Chinese market. It was a good opportunity to closely observe a

Table 1 Overview of the research

Research method	Objects of the research			
Face-to-face interview	Knitwear industry		Design education	
	Chinese knitwear industry	British knitwear industry	Design students	Design teaching staff
	A total of 21 interview (with 9 designers, 1 company owner, 1 product manager, 4 marketing managers, 3 sale assistants and 3 technique managers from 6 Chinese knitwear brands and 2 design studios)	1 designer interview	10 interviews with 13 students from China, UK, and Italy	14 interviews with 14 teachers from UK and China
Case study	Cooperated with 1 Chinese knitwear company for 16 days			
Design practice research	Cooperated with 1 Chinese knitwear company for 45 days, to design a collection with 23 outfits (55 piece of knitwear), to collect feedback throughout exhibitions			
	Witnessed the design process from inside the cooperated knitwear company			
Online survey 1	Female consumer lifestyles and consumption habits in knitwear			
	Survey in China		Survey in UK	
	450 responses		133 responses	
Online survey 2	Female knitwear (cashmere) design and style preference survey			
	Survey in China			
	299 responses			

representative Chinese knitwear brand; the data were helpful to understand the Chinese knitwear brands, in terms of history, development processes, advantage and shortcomings, their competitiveness and their dilemmas.

To understand Chinese knitwear consumers and their preferences of knitwear design and consumption, two online surveys with Chinese and Western knitwear consumers were also undertaken in 2015 and 2016. A total of 583 responses were collected in the first survey consisting of 450 Chinese and 133 international consumers. The survey was designed for collecting Chinese female consumers dressing habits in knitwear consumption. Eighteen questions were divided into three main parts which include dressing habits, consumption habit preferences, lifestyle, and personal background. The survey included both a Chinese and English version. The Chinese version used the online survey Web site called WJX.CN to collect feedback, and the English version used Survey Monkey. In the second survey, 299 responses from women were collected in China with 78.92% (236) people from 25 to 45 age groups. This survey was designed for collect consumers purchasing habits and expectation in cashmere knitwear. The survey consisted of 27 questions with four main parts which included shopping style and habits, dressing style and preferences, cashmere knowledge and brand preferences, and personal background and lifestyles. It used a same WJX.CN Web site as the first survey for Chinese consumers' responses. Both of these two Chinese surveys were linked to the social media WeChat to help with the response rate. Limited by the diffusion path and social circle, 93.34% of responders were young female urban residents which consisted of 21.11% under 25 years old, 41.56% from 25 to 35 age group, and 30.67% from 35 to 45 age group. The English version was linked with Facebook to ask people helping to fill in the survey. Both of the surveys used the online survey Web sites' tools for the result data analysis. These two online surveys helped to understand the different habits between Chinese consumers and international consumers in knitwear purchasing. In the second survey, the data helped to understand what Chinese female consumers expected in cashmere knitwear products.

Design practice research was a significant part for the whole research. Based on the first author's personal work experiences in the Chinese knitwear industry and making use of the research data of the market problems, a knitwear collection for the Chinese market was designed. To explore a way for Chinese knitwear brands to work in design and production, 21 outfits (55 pieces) which combined modern western styles with traditional Chinese themes were designed. Feedback on the designs from the knitwear industry, academics, and other groups in both the West and China was collected. The design practice research was a cooperation project with the case study company, a Chinese knitwear brand. The first author stayed over two months in the company to create the collection in late 2016. In the process of making the collection, she could observe the whole design process both of her design project and the other designers in the company. This provided useful insights into the company's design culture and management. The design collection was displayed in the cooperating company after completion. The knitwear was modeled at an art and design fashion show in Shenzhen with Tsinghua University at the end of 2016. The collection was exhibited at the designer fashion show in Birmingham Fashion Festival (UK). Finally, a three-day exhibition was held in Birmingham City University in 2017. The research data collected from the methodology was analyzed using statistical methods for quantitative data interpretation as well as qualitative analysis of interviews and feedback.

The Challenges of the Chinese Knitwear Industry

Most Chinese knitwear companies began their businesses as knitting manufacturers, but now also operate as suppliers for international brands. Many Chinese knitwear companies offer top quality knitwear for luxury brands, but are unable to run their own brands. For example, during the design practice period, the case study company made beautiful high-quality cashmere knitwear for Japanese and European luxury brands. However, its own knitwear brand looks less attractive although it has the same high qualities as the luxury brands. This shows that Chinese knitwear brands do not lack good manufacturing capabilities. Their problems lie with their branding and design.

Based on the research and the observations, it showed that the gap between successful international brands and Chinese local brands lies in the following aspects:

- A clear brand identity;
- Fashionable designs and images;
- Comprehensiveness of the product range;
- Bigger shopping areas;
- Better locations of the stores;
- More influences on young people;
- Clear brand positioning.

Successful international brands tend to have a clear brand identity with new and appealing products every season. Luxury brands have products with unique designs, good materials and qualities and offer good after-sale services with shops in the best locations. Most luxury brands offer "life warranties, flexible return polices, free adjustments and alterations, free delivery, personalization of products, customization, VIP loyalty programs, personal shopping assistance in stores, special collection viewing and updates" (Hameide, 2011). Many luxury brands successfully build their fashionable and ideal lifestyles to meet consumers' needs. According to Hameide (2011) "a lifestyle brand successfully identifies itself with a lifestyle and a marketing segment to the point that its name or image is mentally triggered when the segment is mentioned." Many global successful fashion brands offer comprehensive product ranges for their consumers, which not only includes clothes, and many types of fashion accessories as well as cosmetics and perfume. This makes the brands attractive for many people, although not everybody can afford the products. In fact, for many young people, aspirational luxury brands not only offer beautiful products, they also become a symbol of success.

While a small percentage of Chinese brands have been successful in the Chinese market, most Chinese brands are not competitive enough with international brands in China. The following observations were made through face-to-face interviews, visits and observations of fashion markets in Beijing, Guangzhou, Hohhot and Baotou cities. While domestic brands aimed for lower price point than luxury brands, they are still affected by international competition. Most global brands occupied better locations with bigger stores than the domestic brands. However, most Chinese brands prefer to follow the market of existing designs. For example, to copy European brands, so that it is difficult for them to build up distinct identities and unique brand personalities. This applied to knitwear brands. In Chinese shopping malls, brands with similar market positioning are clustered and the best positions are given to the most popular brands. For instance, in many shopping malls and department stores, the ground floors (or first floors) are given to international famous brands. The bigger and well-known Chinese womenswear brands are located on the second and third floors, and then the menswear brands on the higher floors. The knitwear brands are always found with other items, such as luggage, outdoor products, and other non-fashion items on the top floors. Most consumers can find clothes they like on the lower levels and do not go to the higher levels. Compared to fashion brands, most knitwear brands have fewer opportunities to be visited by shoppers. In addition, most knitwear brands just offer knitwear for autumn and winter and rarely have a range of matching products. In spring and summer, most of them just sell discounted winter clothes, which does not incentivize the shopping malls to give them prominent locations.

Reasons for the Problems of the Chinese Knitwear Brands

The results of this study suggested the following reasons for the current status of the Chinese knitwear industry. Firstly, the designs from Chinese knitwear brands are not strong enough to support the brands' continual development and survival. The quantity of discounted or unsold stock from these knitwear brands clearly shows that their knitwear designs are not attractive enough for their targeted consumers. Secondly, the competition in the Chinese retail markets, after thousands of local brands and international brands (because most of fashion brands offer knitwear as well) established themselves, is heating up. The third important aspect is Chinese consumers' expectation are evolving quickly, which is influenced by globalization and Internet access. Most Chinese consumers follow the leading fashion trends when they are shopping (Euromonitor International, 2013).

Missing Brand Distinction

The stock of Chinese knitwear brands in retail stores looks the same from year to year and between brands. It is hard to distinguish between brands without looking at their trademarks. For most Chinese knitwear brands, there is no clear differentiation into different markets. The products lack distinctive styles for markets, and the companies do not have clear approaches to build up their brands. For example, the case study brand did not have clear themes and color tones for their autumn and winter products in 2016. There were over 200 knitwear tops with common styles in multiple colors without a clear structure to the product offering in their showroom during a June 2016 visit. Through discussions with the company marketing managers and designers, it became clear that their commercial agents and clients just order products following their preferences. The company did not offer guidance for product structure and order lists of their products in the seasonal purchasing meetings. Thus, it is hard to show distinctive styles and brand images to consumers in retail shops when the company lacks clear design plans and brand strategies.

According to Hameide (2011), "a successful brand has a strong identity (mentally and physically), is innovative, consistent, competitively positioned, and holds a matching positive image in the consumer's mind." Successful western fashion brands offer creative collections with new themes and colors with their brands' DNA every season. Normally, different brands have their own product structures based on the brand positioning and features. Before they begin a new collection, design directors need to make a design plan with themes, colors, and the purpose of the design details to guild their design teams in the product development processes. "A collection is not a random selection of great items. It must have an overall direction, and the piece should work together as a group" (Gehlhar, 2008). In the collection, "the pieces should be able to be worn together or fit in as accent items that can be sold alone" (Gehlhar, 2008).

Dynamic Market Structure in China

The Chinese market is dynamic and the purchasing power is still strong, especially in the first- and second-tier cities. There are some unavoidable challenges, which affects all brands, whether luxury or cheap, international or local, fast fashion or slow fashion:

- Internet shopping: Many people no longer go to shops to buy products, including clothing and fashion items. Online shopping is the first choice for many Chinese consumers.
- Discounts: Knitwear brands keep discount areas in their retail stores throughout the year, and consumers do not pay the full price for the products.
- Copies: Due to the immature intellectual property protection system and social environment, original designs and smaller brands are losing out because of counterfeiting. In the knitwear market, small manufacturers copy designers' and big brands' designs, offering a lower price, and wholesale brands can offer the same products with lower prices for the consumers.

Missing Design Culture in Companies

"Design culture as engaging the inter-relationships of domains of designers, production, consumption with the design object, image, or space," consists of value, circulation, and practice (Julier, 2006). "Design culture was to explore how each of these nodes affected the other—how these relationships were materialized through practices in the design profession and of production and consumption" (Julier, 2006). From another position, "Design culture is located in communication, it is both something designers do, but also is something that is all around" (Julier, 2006). Similarly, design culture for the Chinese knitwear brands, should include: design thinking, the comprehension of how designers think and work (Cross, 2011), design understanding, design process and management, which relates to the "designer's role in the creation of value" (Julier, 2006), design education, and design abilities. There is an absence of a creative design culture and design understanding in the Chinese knitwear industry. Generally, there is a lack of "value creation" and "less focus on innovation" in many companies in China (Roll, 2005). For most Chinese knitwear brands, their design culture is missing the following aspects: clearly defined design processes, valuing designer, thought through product structure and design understanding. In the research processes, the author found many companies did not make clear design plans for their new collection. The designers take their new design inspiration from existing market design samples or photographs from fashion magazines rather than developing a design plan with themes based on fashion trends analysis. Most designers are not encouraged to make innovative products. In most companies, the designers are assessed by how many of their designs are ordered. Many company leaders do not really think designers should play a key role for their brands, so designers are managed in the same way as production line factory workers. Thus, less attractive products cause many woollen knitwear brands to lose their market share. While the cashmere knitwear brands have similar problems to the woolen brands in designing and branding, they benefit from the fact that cashmere is too expensive to use in many fashion brands. Therefore, some big cashmere brands still have space in the market for good-quality cashmere products.

Many knitwear companies pay little attention to developing their design abilities and brand strategies. This has led to some of them going out of business. For example, 5 of 9 Chinese knitwear brands contacted for this research no longer had their own designers. They had given up on creating and producing new products for their own brands by replacing them with purchasing finished knitwear garments in wholesale markets and then inserting their own label into the product. This reflects a lack of understanding of what design should do for a brand, and a belief that design is not important for the competitiveness of a brand. Another reason for not having their own design team is the lack of well-trained knitwear designers. For example, when the author interviewed a knitwear company owner in 2016, he explained that designers were too expensive to employ for his brand. In the late 1990s, his company had designers and around 60 retail stores, but he gradually lost this business and closed all stores until only two were left in Hohhot city in 2016. He believed buying knitwear to put in his stores from wholesale market was an effective way to reduce market risks and inventory pressure. In his experience, the designers who were working for his company were not helpful to in developing the business.

Missing Design Education in Chinese Universities

The ability to design knitwear is related to design education and training at universities and in companies. Compared to fashion design in British higher education, China has completely different teaching methods and concepts. Fashion courses aim to teach systematic knowledge in fashion and textiles to their students. For example, they cover basic courses such as pattern cutting, sewing skills, fashion illustrator, and so on. They have a strong basis in theory with little practice. Students must study and pass around 30 subjects, such as fashion history, fashion technique skills, color printing, sketch, color composition, in their first three years, before undertaking a project in their last year of degree study. In contrast, the western fashion education system trains their students to build a design approach with research and development, which enables them to develop inspirational design ideas for clothes. For example, in many universities in the UK, students are required to do many different projects in their three years of study, so that students can develop a personal system for their own styles and preferences. Many British designers have really strong design abilities when they graduate. However, the majority of new Chinese designers require a long time to adapt to the design processes and systems in their first employment position. Many companies are skeptical of the professional skills of new designers, and do not give them enough space and freedom as part of the design processes. On the other hand, the knitwear companies do not train new designers themselves. This makes it difficult for many designers to develop their design abilities once they work for companies. Most designers in Chinese knitwear brands do not have the skill or confidence to create innovative designs. Therefore, they follow existing markets rather than creating different products with distinctive designs for their brands.

Lack of Understanding of the Knitwear Consumers

Knitwear occupies a significant share of the fashion market. Most fashion brands offer knitwear to match their other clothing ranges and to improve their sale performances. In recent years, living conditions and personal transportations have improved in China, so that many consumers no longer have to rely on knitwear to keep warm in winter. Thus, the knitwear market had to respond to the customers' demands for aesthetically appealing, rather than simply warming, clothing. According to the online surveys of Chinese female consumers on knitwear consumption and preferences of cashmere knitwear design, almost half (49.56%) of 450 respondents claimed to make purchasing decisions based on design and styles when buying knitwear, 38.44% of responders bought knitwear for the materials and qualities and another 12% consumers make decisions based on knitwear colors (see Fig. 1).

Furthermore, in a second survey (299 participants), where customers were asked to rank their preferences, design also occupied the first place among the aspects that influence the purchasing decisions for cashmere knitwear (see Fig. 2).

Recommendation for Changing the Situation

In order to help Chinese knitwear brands to deal with their problems, it is strongly recommended that the following seven points are addressed:

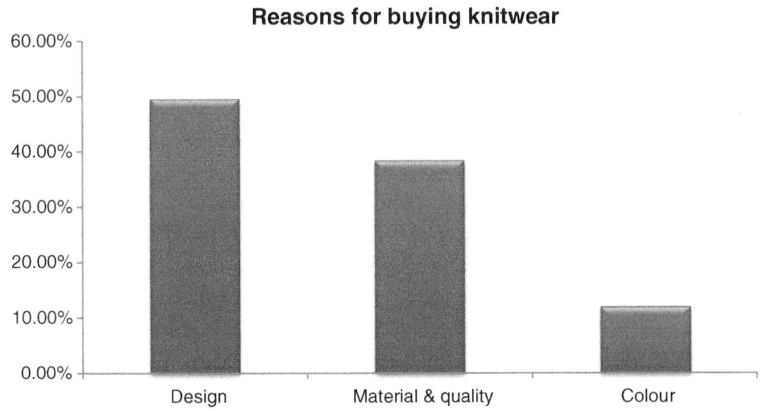

Fig. 1 Main reasons for buying new knitwear

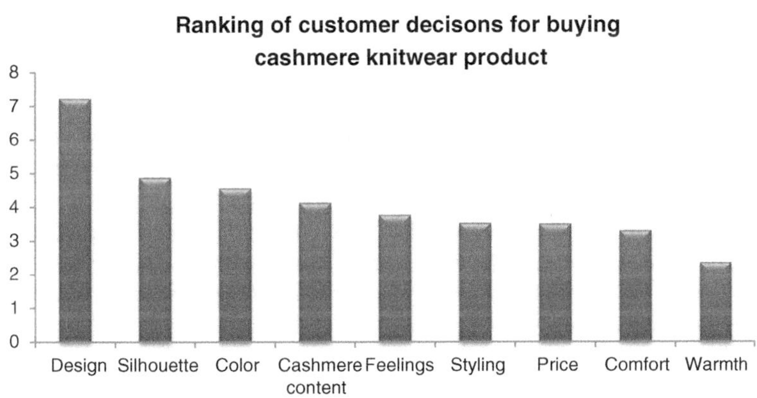

Fig. 2 Ranking of main factors of purchasing decisions for cashmere knitwear survey

1. Hire young designers who are western trained or recently graduated;
2. Give designers time, money, and space to design;
3. Commit to strategic investment in design over many years;
4. Clarity brands positioning and market segments;
5. Define a target customer for the brand;
6. Offer complete series collections with other products and knitwear;
7. Invest in building up a brand: stores, marketing, advertising.

Hiring young designers can bring new ideas to the brands; many younger designers with western education background or taught by western background teachers may bring new design methods for the old design teams. At the same time, give more time and space for designers, to encourage them to create different designs to gradually build the brand identity. Building successful brands needs improvement in many aspects. Spending more time to build their own design team, investment in designers' training and other opportunities to update their knowledge and skills should be long-term development goals for brands. Clarify the brands positioning and market segments to understand who are the target consumers and what are their lifestyles and real demands are. Completing the design collections, create other products with knitwear to satisfy consumer needs. Building the brands image through designing and decorating retail stores, developing marketing strategies and suitable advertisings, as well as, taking the brands to fashion shows or other platforms is essential. Furthermore, building Web sites and other channels to close the distance from consumers to the brands are highly desirable. Achieving all of these will need investment and time. For many companies, the first step should to re-build the brands and re-design their products.

Conclusion

In conclusion, China is still a really dynamic fashion market with a population of over 1.4 billion and a steadily growing economy. This can still be explored and developed with potential opportunities for the fashion markets, including knitwear brands. As a fast-growing developing country, there are many problems that are unavoidable in the fashion market. Local brands were built quickly and expanded rapidly, which was bound to cause many problems like supply-demand imbalance and brand homogenizations. However, both the consumers' lifestyles and purchasing habits are evolving quickly, and the local governments are trying to gradually improve management and laws for the fashion markets. Influenced by the continually growing economy, Chinese fashion consumption power should still have a huge potential for growth in the future. After the rapid development at the beginning of 1990s and 2000s, the Chinese knitwear industries are suffering from many problems including the missing brand distinctions, the missing design cultures, the missing design processes, and the missing understanding of consumers. Thus, re-orientation and re-design of their brands and products are a necessary way for development and survival. Knitwear brands need to pay more attention to identify their market sections and brand strategies to try to build their brands with distinct characteristics, through improved design culture and awareness. There are still opportunities to grow their brands and win back the markets and their target consumer groups in China and beyond. In the academic aspects, Chinese design universities have to address the imbalance between educational outcomes and the real needs of industry. They must offer more design practice and design thinking models that are necessary for the development of students' design abilities. This can both enhance new designers' employment and their confidence. There are few researchers who focus on knitwear research in the world until now. Thus, for knitwear designers, industry, companies, or related researchers, there are limited materials and related theory can be found to support their working, training, studying, and researching. The methodology and findings of this research provide a rigorous approach to fill the gaps between Chinese knitwear, design, and consumers. The work offers fundamental research, based on sound theoretical methods, to provide future framework opportunities to improve the status of the Chinese knitwear sector.

References

Anon. (2008, June 23). Chinese cashmere brands white paper. *Textile Apparel Weekly*, pp. 50–53.

Anon. (2009). Survey about fashion brands' market share in Beijing city in 2008. *Commercial Medal* (pp. 62–67). China: Beijing Business Information Center.

Anon. (2013). The road of transformation of the department stores. *China Business Trade, 9*, 6–8.

Anon. (2016). Statistics. *Journal of Chinese Costume, 3*, 13.

Anon. (2017). *39 college students were sued college students bear high loans to buy luxury good break faith, whose fault*? Retrieved from: http://jiangsu.china.com.cn/html/jsnews/society/10509391_1.html.

Chen, T. H. (2011). Face consciousness and female brand clothing consumer marketing strategy. *International Conference on Engineering Business Management*, 22–24 Mar 2011, Wuhan, China.

Chen, W. W. (2013). There is excess in the stock street. *Zhejiang Daily, 16*(9), 9.

Cross, N. (2011). *Design Thinking*. Oxford, Berg Publishers.

Euromonitor International. (2013). *Consumer lifestyles in China*. Retrieved from: www.euromonitor.com.

Euromonitor International. (2017). *Consumer lifestyles in China*. Retrieved from: http://www.euromonitor.com/consumer-lifestyles-in-china/report.

Gehlhar, M. (2008). *The fashion designer survival guide*. New York: Kaplan Publishing.

Hameide, K. K. (2011). *Fashion branding unraveled*. New York: Fairchild Books, A Division of Conde Nast Publication.

Julier, G. (2006). From visual culture to design culture. *Design Issues, 22*(1), 64–76.

Lao, Y. M. (2013). Research on clothing consumption habits of older consumer-a case study of Shaoxing Region. *China Textile Leader, 12*, 89–91.

Li, X. Y. (2004). *A study of the basic Development of fashion marketing in China*. M.A. Thesis, Dong Huan University, China.

Lia, K. K., & Zaichkowsky, J. L. (1999). Brand imitation: Do the Chinese have different views? *Asia Pacific Journal of Management, 16*, 179–192.

Liang, J., & Wang, Y. (2007). Reflection on excessive consumption. *Science Technology and Industry, 7*(5), 36–38.

Lin, L. (2012). Traditional department store industry in their cold current. *Economic Life, 11*, 272.

Li, G. X., Li, G. F., & Kambele, Z. (2012). Luxury fashion brand consumers in China: Perceived value, fashion lifestyle, and willingness to pay. *Journal of Business Research, 65*, 1516–1522.

Liu, B. R. (2007, January 19). Demonic market is surplus and clothing brands with aggravate competition. *Information Daily*, p. the 2nd section, 1.

Qin. (2012, December 18). Popular youth consumption trend in Chinese luxury market. *China Industrial and Commercial Daily*, p. B04.

Roll, M. (2005). *Asian brand strategy: How Asia builds strong brands*. New York, NY: Palgrave Macmillan.

Situ, W. C. (2015). The department stores continually are closing: The in-stores future. *Economy, 6*, 58–61.

Sun, J. H. (2014). *A research on brand effect of luxury brand and Chinese luxury consumers*. M.I.B. Thesis, Nanjing University, China.

Tsui, C. (2010). *China fashion: Conversations with designers*. Oxford: Berg Publisher.

Wang, L. (2010). Psychoanalysis for consumption of luxury clothing products. *Garment Art, 12* (04), 90–91.

Wang, R. R. (2014). *Empirical Research between consumers' lifestyles and clothing purchase behaviors of "after 70s-80s-90s"*. M.A. Thesis, Zhejiang Sci-Tech University, China.

Xu, F. L. (2015). Excessive consumption and happiness alienation. *Journal of Shenyang Normal University (social science edition), 39*(187), 26–28.

Xu, Y. Y. (2013). The influence of Chinese female consumers lifestyles on Korean fashion brands attitudes and loyalty. *Eighteenth Industrial Design Annual Conference Collected Papers*, pp. 437–442.

Yang, D. J. (2009). "Soft gold" big brands rise in the time: thoughts and suggestions for Chinese cashmere industry. *Textile and Apparel Weekly, 2009*(2), 70–71.

Ye, J. L. (2013). Research on overseas consumption in the dress of Shanghai consumers. M.A. Thesis, Donghua University, China.

Yu, S. Y. (2015). The study of clothing consumption and dressing collection about middle-aged women. M.A. Thesis, Shanxi University of Science and Technology, China.

Zhao, J. R. (2009). Contemporary university students' consumption behaviors. *Contemporary Economic, 1*(2), 114–116.

Zhang, J. (2015). Understanding the concept of shame in the Chinese culture. *NYS Child Welfare/ Child Protective Services Training Institute, 4*.

Zhang, Z. (2016). The greatest powers needed for China's garment industry development. *China Fiber Inspection, 11*, 118–119.

Printed in the United States
By Bookmasters